Navigating the ELPS in the Math Classroom

Using the New Standards to Improve Instruction for English Learners

John Seidlitz
C. Araceli Avila

Seidlitz Education
10864 Gulfdale Street
San Antonio, TX 78216

ISBN 978-0-9822078-1-9

For related titles and support materials visit www.seidlitzeducation.com.

Table of Contents

English Learners in Texas and the New Proficiency Standards **5**

How to Use this Manual **6**

Introduction (subsection a) **9**

District Responsibilities (subsection b) **15**

- *Understanding the ELPS Framework* *18*
- *ELPS District Implementation Checklist* *20*
- *ELPS Aligned Walk-Through Observation* *22*

Cross-Curricular Student Expectations (subsection c) **25**

- *ELPS Integration Plan for Teachers* *32*
- *Seven Steps to Builidng a Language-Rich Interactive Classroom* *33*
- *Language Objectives Aligned to Student Expectations* *34*
- *ELPS Lesson Plan Template* *36*
- *Elps Lesson Plan Activity Guide* *41*
- *Engaging Activities that Promote Language Development*
    - Comparing Whole Numbers in a Conga Line *42*
    - Multiple Representations Card Game *45*
    - Order It Up Math Puzzle *51*
    - Parallel and Perpendicular Lines Sort *63*
    - Making Inferences from Graphs *66*
- *Sentence Stems and Activities Aligned to Student Expectations* *71*

Language Proficiency Level Descriptors (subsection d) **111**

Guide to Terms and Activities **129**

Bibliography **141**

This page is intentionally left blank.

English Learners in Texas and the New Proficiency Standards

Research into what works for English language learners indicates that one of the keys to success is a consistent focus on content area language acquisition (Gibbons, P., 2002; Samway, K., 2006; Echevarria, Vogt, Short, 2008; Zweirs, J. 2008). This approach emphasizes the need to intentionally make content comprehensible while developing academic language skills for students acquiring English as a second language. It also requires academic language instruction be integrated into all areas of instruction so that all teachers of ELLs understand their role as a teacher of both content and language.

In 1998 the second language acquisition (ESL) standards were adopted as part of the Texas Essential Knowledge and Skills for Spanish Language Arts and English Language Arts. These standards included student expectations and descriptions of proficiency levels for reading, writing, listening, and speaking for students at various levels of language proficiency. Because these standards were only integrated into English and Spanish language arts, few content area teachers made use of them when planning instruction for English learners.

In 2006 the Texas Education Agency assembled a group to review and revise the ESL standards. The result was the new English Language Proficiency Standards that went into effect in December of 2007. The standards describe the instruction that districts need to provide for English language learners so that they can successfully master both content area knowledge and academic language. Unlike the former ESL standards, the new standards clearly indicate that the ELPS are to be integrated into *each subject of the required curriculum.* The new ELPS contain a brief introduction into what kind of instruction is required for English learners, an outline of district responsibilities, cross curricular student expectations, and language proficiency level descriptors.

Although integrating the ELPS into content areas will take time and effort for Texas educators, the new standards provide a unique opportunity to improve instruction for English learners. The purpose of this manual is to provide resources and tools that will help make the process of implementation easier for administrators, specialists, and teachers who serve English learners in classrooms across Texas.

How to Use this Manual

This manual is designed to help teachers, administrators and specialists use the ELPS (*http://www.tea.state.tx.us/rules/tac/chapter074/ch074a.html*) to improve instruction for English learners. It is divided into four sections corresponding to the four sections of the English Language Proficiency Standards:

(a) Introduction
(b) District responsibilities
(c) Cross-curricular student expectations
(d) Language proficiency level descriptors

Each section begins with the text of the subsection of chapter 74.4 (TAC) that the tools and resources will address. This is followed by an assessment of current levels of understanding and implementation. The assessment is designed to guide administrators, educators and specialists through a process of deciding which tools and resources will best meet their current needs. The assessment is followed by tools for understanding the meaning of the document as well as templates for implementing changes at the classroom, campus, and district levels in line with the new standards.

Section (a) addresses the introduction to the *English Language Proficiency Standards (2008).* The section includes summaries of: the introduction (subsection a), district responsibilities (subsection b), the student expectations (subsection c) and the language proficiency level descriptors (subsection d). The summaries are followed by a handout listing reasons why the ELPS are significant to the instruction of English learners in Texas.

Section (b) outlines district responsibilities discussed in the ELPS. The first set of tools is designed to help educators understand the ELPS framework for English proficiency. They help clarify specific terms found in chapter 74.4. The terms ***communicated, sequenced*** and ***scaffolded*** that describe ***linguistically accommodated instruction*** and the terms ***focused, targeted, and systematic*** that describe high quality second language acquisition instruction are discussed. The next set of tools includes an implementation checklist and a series of planning, observation, and coaching tools that can be used to meet the needs of teachers and administrators implementing ELPS at the classroom or campus level.

Section (c) addresses cross-curricular student expectations. The tools are designed to help educators integrate the ELPS into content area instruction. A quick guide to reading the new standards is included as well as specific lesson planning templates and sample lessons aligned to the ELPS. An ELPS Integration Plan for Teachers walks educators through a process for creating lesson plans that target academic language and concept development. Another tool outlines a seven step process for using the ELPS to create a language rich interactive classroom. The last tool in this section is a comprehensive guide for planning instruction using the ELPS that includes activities and suggested sentence stems that can be used with each of the standards.

Section (d) contains tools and resources for planning instruction based on students' language proficiency levels. A summary of each language level and specific strategies corresponding to each proficiency level as well as tools for differentiating instruction are provided. The section does not include tools for formal assessment of English learners nor linguistic accommodations for English learners taking the Linguistically Accommodated Testing (LAT). These can be found at the Texas Education Agency website here: http://portals.tea.state.tx.us/page.aspx?id=568.

Guide to Terms and Activities: This section contains a description of each of the activities and strategies mentioned in the manual as well as references for further research on the activity.

This page is intentionally left blank.

Introduction

(subsection a)

Chapter 74. Curriculum Requirements: Subchapter A. Required Curriculum

§74.4. English Language Proficiency Standards

http://www.tea.state.tx.us/rules/tac/chapter074/ch074a.html.

(a) Introduction.

(1) The English language proficiency standards in this section outline English language proficiency level descriptors and student expectations for English language learners (ELLs). School districts shall implement this section as an integral part of each subject in the required curriculum. The English language proficiency standards are to be published along with the Texas Essential Knowledge and Skills (TEKS) for each subject in the required curriculum.

(2) In order for ELLs to be successful, they must acquire both social and academic language proficiency in English. Social language proficiency in English consists of the English needed for daily social interactions. Academic language proficiency consists of the English needed to think critically, understand and learn new concepts, process complex academic material, and interact and communicate in English academic settings.

(3) Classroom instruction that effectively integrates second language acquisition with quality content area instruction ensures that ELLs acquire social and academic language proficiency in English, learn the knowledge and skills in the TEKS, and reach their full academic potential.

(4) Effective instruction in second language acquisition involves giving ELLs opportunities to listen, speak, read, and write at their current levels of English development while gradually increasing the linguistic complexity of the English they read and hear, and are expected to speak and write.

(5) The cross-curricular second language acquisition skills in subsection (c) of this section apply to ELLs in Kindergarten-Grade 12.

(6) The English language proficiency levels of beginning, intermediate, advanced, and advanced high are not grade-specific. ELLs may exhibit different proficiency levels within the language domains of listening, speaking, reading, and writing. The proficiency level descriptors outlined in subsection (d) of this section show the progression of second language acquisition from one proficiency level to the next and serve as a road map to help content area teachers instruct ELLs commensurate with students' linguistic needs.

ELPS Awareness Self-Assessment

Rate the current level of awareness of the English Language Proficiency standards at your district or campus.

A: Always
M: Mostly

S: Sometimes
N: Never

Indicator	A	M	S	N	Comments/Questions
Teachers of ELLs are aware of the unique needs of ELLs to acquire social language skills for social interaction.					
Teachers of ELLs are aware of the unique needs of ELLs to acquire academic language necessary for academic tasks.					
Teachers of ELLs are aware of the need to integrate language and content instruction to ensure that ELLs acquire social and academic English.					
Teachers of ELLs receive staff development on the unique needs of ELLs to acquire social and academic English.					
ELLs are assessed for their language proficiency level upon entry into the district.					
Teachers of ELLs are aware of the language levels of their students on the initial district assessment and on the TELPAS.					
Administrators, teachers and specialists are aware of the need to integrate the cross-curricular student expectations of the ELPS into content area instruction.					

Summaries of ELPS*
Introduction, District Responsibilities and Student Expectations
(subsection a,b,c)

Introduction	District Responsibilities
a1: Part of required curriculum for each subject including **proficiency standards** and **level descriptors** a2: ELLs need social and academic English language proficiency to be successful a3: Instruction must integrate **social and academic English** in content areas a4: ELLs must read, write, listen, and speak in increasing complexity a5: Student Expectations of ELPS apply to K-12 students a6: Level descriptors are not grade specific and serve as a road map.	b1: Identify students' proficiency levels using proficiency level descriptors b2: Provide **linguistically accommodated** content instruction (communicated, sequenced, scaffolded) b3: Provide linguistically accommodated content-based language instruction b4: Focused, targeted, and systematic language instruction for beginning and intermediate ELLs (Grade 3 or higher)

Learning Strategies	
c1A: Use prior knowledge to learn new language c1B: Monitor language with self-corrective techniques c1C: Use techniques to learn new vocabulary c1D: Speak using learning strategies	c1E: Use and reuse new basic and academic language to internalize language c1F: Use accessible language to learn new language c1G: Distinguish formal and informal English c1H: Expand repertoire of language learning strategies

Listening	Speaking
*c2A: Distinguish sound and intonation c2B: Recognize English sound system in new vocabulary c2C: Learn new language heard in classroom interactions and instruction c2D: Monitor understanding and seek clarification c2E: Use visual, contextual linguistic support to confirm and enhance understanding c2F: Derive meaning from a variety of media c2G: Understand general meaning, main points, and details c2H: Understand implicit ideas and information c2I: Demonstrate listening comprehension	c3A: Practice using English sound system in new vocabulary c3B: Use new vocabulary in stories, descriptions, and classroom communication c3C: Speak using a variety of sentence structures c3D: Speak using grade level content area vocabulary in context c3E: Share in cooperative groups c3F: Ask and give information using high-frequency and content area vocabulary c3G: Express opinions, ideas and feelings c3H: Narrate, describe and explain c3I: Adapt spoken language for formal and informal purposes c3J: Respond orally to information from a variety of media sources

Reading	Writing
*c4A: Learn relationships of sounds and letters in English *c4B: Recognize directionality of English text c4C: Develop sight vocabulary and language structures c4D: Use prereading supports c4E: Read linguistically accommodated content area materials c4F: Use visual and contextual supports to read text c4G: Show comprehension of English text individually and in groups c4H: Read silently with comprehension c4I: Show comprehension through basic reading skills c4J: Show comprehension through inferential skills c4K: Show comprehension through analytical skills	*c5A: Learn relationships between sounds and letters when writing c5B: Write using newly acquired vocabulary c5C: Spell familiar English words c5D: Edit writing c5E: Employ complex grammatical structures c5F: Write using variety of sentence structures and words c5G: Narrate, describe, and explain in writing

* *These summaries must be used in conjunction with cross-curricular student expectations when planning instruction.*

* only with new readers

Summaries of ELPS: Proficiency Level Descriptors*

(subsection d)

Level	Listening *(d1: k-12)* *The student comprehends...*	Speaking *(d2: k-12)* *The student speaks...*	Reading *(d4: 2-12)* *The student reads...*	Writing *(d6: 2-12)* *The student writes...*
Beginning (A)	1A(i) few **simple conversations** with linguistic support 1A(ii) **modified conversation** 1A(iii) few words, **does not seek clarification,** watches others for cues	2A(i) using **single words and short phrases** with practiced material; tends to give up on attempts 2A(ii) using **limited bank of key vocabulary** 2A(iii) with **recently practiced familiar material** 2A(iv) with frequent **errors** that hinder communication 2A(v) with **pronunciation that inhibits communication**	4A(i) little except recently practiced terms, **environmental print,** high frequency words, **concrete words represented by pictures** 4A(ii) **slowly, word by word** 4A(iii) with very limited sense of English structure 4A(iv) with comprehension of **practiced, familiar text** 4A(v) with need for **visuals and prior knowledge** 4A(vi) modified and **adapted text**	6A(i) with **little ability to use English** 64A(ii) **without focus** and coherence, conventions, organization, voice 6A(iii) labels, lists, and copies of printed text and **high-frequency words/phrases**, short and simple, practiced sentences primarily in **present tense with frequent errors** that hinder or prevent understanding
Intermediate (B)	1B(i) unfamiliar language with **linguistic supports** and adaptations 1B(ii) unmodified conversation with **key words** and phrases 1B(iii) with **requests for clarification** by asking speaker to repeat, slow down, or rephrase speech	2B(i) **with simple messages** and hesitation to think about meaning 2B(ii) using **basic vocabulary** 2B(iii) with **simple sentence structures** and present tense 2B(iv) with **errors** that inhibit unfamiliar communication 2B(v) with **pronunciation generally understood** by those familiar with English language learners	4B(i) **wider range of topics:** and everyday academic language 4B(ii) **slowly** and **rereads** 4B(iii) basic language structures 4B(iv) simple sentences **with visual cues, pretaught vocabulary** and **interaction** 4B(v) **grade-level texts** with difficulty 4B(vi) at high level with **linguistic accommodation**	6B(i) with **limited ability to use English** in content area writing 6B(ii) best on **topics that are highly familiar** with simple English 6B(iii) with **simple oral tone in messages**, high-frequency vocabulary, loosely connected text, repetition of ideas, **mostly in the present tense**, undetailed descriptions, and **frequent errors**
Advanced *(C)*	1C(I) with some processing time, **visuals, verbal cues, and gestures; for unfamiliar conversations** 1C(ii) most unmodified interaction 1C(iii) with occasional **requests** for the speaker to slow down, repeat, rephrase, and **clarify meaning**	2C(i) in conversations with some **pauses to restate, repeat, and clarify** 2C(ii) using **content-based and abstract terms** on familiar topics 2C(iii) with **past, present, and future** 2C(iv) using **complex sentences** and grammar with some errors 2C(v) with **pronunciation usually understood** by most	4C(i) abstract grade appropriate text 4C(ii) **longer phrases and familiar sentences** appropriately 4C(iii) while developing the ability to construct meaning from text 4C(iv) at **high comprehension** level with **linguistic support** for unfamiliar topics and to clarify meaning	6C(i) grade appropriate **ideas with second language support** 6C(ii) with extra need for second language **support when topics are technical and abstract** 6C(iii) with a grasp of basic English usage and some understanding of complex usage with **emerging grade-appropriate vocabulary** and a more academic tone
Advanced High *(D)*	1D(i) longer **discussions on unfamiliar topics** 1D(ii) spoken information nearly **comparable to native speaker** 1D(iii) with few requests for speaker to slow down, repeat, or rephrase	2D(i) in **extended discussions** with few pauses 2D(ii) using **abstract content-based vocabulary** except low frequency terms; using idioms 2D(iii) with grammar **nearly comparable to native speaker** 2D(iv) with **few errors** blocking communication 2D(v) **occasional mispronunciation**	4D(i) **nearly comparable to native speakers** 4D(ii) **grade appropriate familiar text** appropriately 4D(iii) while constructing meaning at near native ability level 4D(iv) with high level comprehension with **minimal linguistic support**	6D(i) grade appropriate **content area ideas with little need for** linguistic **support** 6D(ii) develop and demonstrate **grade appropriate** writing 6D (iii) nearly **comparable to native speakers** with clarity and precision, with **occasional difficulties** with naturalness of language.

***These summaries are not appropriate to use in formally identifying student proficiency levels for TELPAS. TELPAS assessment and training materials are provided by the Texas Education Agency Student Assessment Division**: http://www.tea.state.tx.us/index3.aspx?id=3300&menu_id3=793*

Why the ELPS?

1. English language learners benefit from content area instruction that is accomodated to their need for comprehensible input (Krashen, 1983; Echevarria, Vogt, and Short, 2008).

2. English language learners benefit from academic language instruction integrated into content area instruction (Crandall, 1987; Snow et. al. 1989).

3. English language leaners benefit from programs that hold high expectations for students for academic success (Collier, 1992; Lucas et al, 1990, Samway & McKeon 2007).

4. Language proficiency standards provide a common framework for integrating language and content instruction for English learners (Short, 2000).

District Responsibilities

(subsection b)

§74.4. English Language Proficiency Standards

http://www.tea.state.tx.us/rules/tac/chapter074/ch074a.html.

(b) School district responsibilities. In fulfilling the requirements of this section, school districts shall:

(1) identify the student's English language proficiency levels in the domains of listening, speaking, reading and writing in accordance with the proficiency level descriptors for the beginning, intermediate, advanced and advanced high levels delineated in subsection (d) of this section;

(2) provide instruction in the knowledge and skills of the foundation and enrichment curriculum in a manner that is linguistically accommodated (communicated, sequenced and scaffolded) commensurate with the student's levels of English language proficiency to ensure that the student learns the knowledge and skills in the required curriculum;

(3) provide content-based instruction including the cross-curricular second language acquisition essential knowledge and skills in subsection (c) of this section in a manner that is linguistically accommodated to help the student acquire English language proficiency; and

(4) provide intensive and ongoing foundational second language acquisition instruction to ELLs in Grade 3 or higher who are at the beginning or intermediate level of English language proficiency in listening, speaking, reading, and/or writing as determined by the state's English language proficiency assessment system. These ELLs require focused, targeted, and systematic second language acquisition instruction to provide them with the foundation of English language vocabulary, grammar, syntax and English mechanics necessary to support content-based instruction and accelerated learning of English.

ELPS Implementation Self-Assessment

Rate the current level of implementation of the English Language Proficiency standards at your district or campus.

A: Always
M: Mostly

S: Sometimes
N: Never

Understanding the ELPS Framework: (1)

Indicator	A	M	S	N	Comments/Questions
Teachers of ELLs receive sufficient training in how to provide ELLs instruction in social and academic English.					
Teachers of ELLs receive sufficient training on how to differentiate instruction based on the language levels of English learners.					
Teachers of ELLs integrate language and content area instruction in their lesson plans.					
Teachers of ELLs provide linguistically accommodated instruction to meet the language proficiency levels of their English learners.					
ELLs have opportunities to read and write in academic English during content area instruction.					
ELLs have opportunities to listen and speak using academic English during content area instruction.					
The cross-curricular student expectations are being integrated into existing curriculum frameworks.					
The cross-curricular student expectations are being integrated into content area lesson plans.					

Understanding the ELPS Framework : (1)
Linguistically Accommodated Instruction

Curriculum for ELLs must be:	What is it?	What are some examples?
Communicated	Comprehensible input is used to convey the meaning of key concepts to students. (Krashen, 1983)	• Visuals, TPR (Total Physical Response) and other techniques to communicate key concepts • Clear explanation of academic tasks • Speech appropriate for language level • Use of Native Language Resources (Echevarria, Vogt, Short, 2008)
Sequenced	Instruction is differentiated to align with the progression of students' language development level. (Hill & Flynn, 2006)	• Differentiating language and content instruction • Targeted use of supplementary materials and resources • Pre-teaching social and academic vocabulary necessary for interaction and classroom tasks (Hill & Flynn, 2006)
Scaffolded	ELLs receive structured support that leads to independent acquisition of language and content knowledge. (Echevarria, Vogt, Short, 2008)	• **Oral scaffolding:** recasting, paraphrasing, wait time, etc. • **Procedural scaffolding:** moving from whole class, to group, to individual tasks. • **Instructional scaffolding:** providing students concrete structures such as sentence and paragraph frames, patterns, and models. (Echevarria, Vogt, & Short, 2008)

Understanding the ELPS Framework: (2)
Foundations of Second Language Acquisition Instruction for Beginning and Intermediate ELLs Grades 3-12

"Make sure the system for second language acquisition instruction focuses on the target."

Second language acquisition instruction must be:	What is it?	What are some examples?
Focused	**Concentrated effort centered on student acquisition** of vocabulary, grammar, syntax and English mechanics necessary to support content-based instruction and accelerated learning of English.	• Explicit instruction in English vocabulary and language structures • Lesson plans include cross-curricular student expectations from the ELPS. • Use of sentence structures of increasing complexity in vocabulary, grammar and syntax.
Targeted	**Specific goals and objectives** align with vocabulary, grammar, syntax and English mechanics necessary to support content-based instruction and accelerated learning of English.	• Content objectives for ELLs align with the TEKS • Language objectives for ELLs align with ELPS and language skills necessary for TEKS • Formal and informal assessments align with content and language assessments.
Systematic	**Well organized structure** in place to ensure students acquire vocabulary, grammar, syntax and English mechanics necessary to support content-based instruction and accelerated learning of English.	• ELPS integrated into district curriculum frameworks • **Comprehensive plan for students in grades 3-12 at beginner or intermediate** level for integrating language and content instruction • Comprehensive plan for assessing the implementation of focused, targeted instruction for beginner and intermediate students in grades 3-12 • Periodic review of progress of ELLs through formal and informal assessment

ELPS District Implementation Checklist

Goal	We will have met this goal when...	Steps	Person(s) Responsible	Dates/Deadlines
Administrators and specialists integrate ELPS into ongoing professional development and evaluation.				
Staff understands the importance of TELPAS and other formal assessments to identify language levels of ELLs.				
Staff understands the need for ELLs to develop social and academic English.				
Staff understands methods for providing linguistically accommodated instruction for ELLs.				
Staff understands cross-curricular student expectations.				
Staff develops a plan for systematic academic language development for ELLs.				
Teachers include ELPS in lesson plans in core content areas.				

Two Key Questions for Assessing Quality Instruction for ELLs

Do English learners understand the key content concepts *(aligned to TEKS)*?	Are English learners developing their ability to read, write, listen and speak in academic English about content concepts *(in ways described in the ELPS)*?

ELPS Aligned Walk-Through Observation

Observer: Class:
Teacher: Date:

Indicator	Comments/Questions
☐ Content and language objectives posted	
☐ Evidence of use of explicit vocabulary instruction	
☐ Evidence of use of variety of techniques to make content comprehensible	
☐ Evidence of reading and writing in academic English	
☐ Evidence of student/student interaction focusing on lesson concepts	
☐ Specific instructional interventions for ELLs appropriate to students' language levels (sentence stems, native language resources, word banks, low risk environment for language production, etc.)	

ELPS Aligned Lesson Observation

Observer: Class:
Teacher: Date:

Indicator	Comments/Questions
☐ Teacher posts and explains clearly defined content objectives aligned to the TEKS to ELLs.	
☐ Teacher posts and explains clearly defined language objectives aligned to the ELPS to ELLs.	
☐ Teacher clearly communicates key concepts, words, phrases and directions for instructional tasks to ELLs (*using visuals, gestures, native language resources, etc. as needed*).	
☐ Teacher differentiates instruction *(alters instruction, language demands and assessment)* to align with the students' language development level.	
☐ Teacher provides verbal and procedural scaffolding for ELLs *(sentence stems, modeling, instruction in strategies etc.)*.	
☐ Teacher provides opportunities for students to read and write using academic English.	
☐ Teacher provides opportunities for ELLs to listen and speak using academic and social English.	
☐ ELLs demonstrate understanding of content and language objectives.	

ELPS Aligned Lesson Observation Coaching Tool

Observer: **Class:**

Teacher: **Date:**

Indicator	Comments/Questions
☐ Teacher posts and explains clearly defined content objectives aligned to the TEKS to ELLs.	• Are the objectives posted? • Do ELLs understand the objectives? • Are the objectives aligned with the TEKS? • Does the lesson align with the objectives?
☐ Teacher posts and explains clearly defined language objectives aligned to the ELPS to ELLs.	• Are the objectives posted? • Do ELLs understand the objectives? • Are the objectives aligned with the ELPS? • Does the lesson align with the objectives?
☐ Teacher clearly communicates key concepts, words, phrases and directions for instructional tasks to English learners *(using visuals, gestures, native language resources, etc. as needed).*	• Do ELLs understand the key concepts? • Does the teacher explicitly teach key concept area vocabulary? • Does the teacher teach ELLs specific words and phrases necessary for instructional tasks? • Do ELLs show a clear understanding of instructional tasks?
☐ Teacher differentiates instruction *(alters instruction, language demands, and assessment)* to align with the students' language development level.	• Is the teacher aware of the students' language levels? • Are instructions, assignments, and assessments appropriate for the students' level of language development?
☐ Teacher provides verbal and procedural scaffolding for ELLs. *(sentence stems, modeling, instruction in strategies etc.)*	• Does the teacher provide models, examples, and structures that enable ELLs to work toward independence? • Do ELLs use specific strategies when they need clarification about content or language?
☐ Teacher provides opportunities for students to read and write using academic English.	• Do ELLs read academic English during the lesson? • Do ELLs write during the lesson? • Are ELLs supported in finding ways to enable them to read and write during the lesson?
☐ Teacher provides opportunities for ELLs to listen and speak using academic and social English.	• Do ELLs listen and speak using social English? • Do ELLs use content area vocabulary during classroom interactions? • Do ELLs use academic English structures during classroom interactions?
☐ ELLs demonstrate understanding of content and language objectives.	• Are ELLs assessed throughout the lesson for understanding of content and language?

Cross-Curricular Student Expectations

(subsection c)

§74.4. English Language Proficiency Standards

http://www.tea.state.tx.us/rules/tac/chapter074/ch074a.html.

(c) Cross-curricular Student Expectations

(1) Cross-curricular second language acquisition/learning strategies. The ELL uses language learning strategies to develop an awareness of his or her own learning processes in all content areas. In order for the ELL to meet grade-level learning expectations across the foundation and enrichment curriculum, all instruction delivered in English must be linguistically accommodated (communicated, sequenced, and scaffolded) commensurate with the student's level of English language proficiency. The student is expected to:

(A) use prior knowledge and experiences to understand meanings in English;

(B) monitor oral and written language production and employ self-corrective techniques or other resources;

(C) use strategic learning techniques such as concept mapping, drawing, memorizing, comparing, contrasting, and reviewing to acquire basic and grade-level vocabulary;

(D) speak using learning strategies such as requesting assistance, employing non-verbal cues, and using synonyms and circumlocution (conveying ideas by defining or describing when exact English words are not known);

(E) internalize new basic and academic language by using and reusing it in meaningful ways in speaking and writing activities that build concept and language attainment;

(F) use accessible language and learn new and essential language in the process;

(G) demonstrate an increasing ability to distinguish between formal and informal English and an increasing knowledge of when to use each one commensurate with grade-level learning expectations; and

(H) develop and expand repertoire of learning strategies such as reasoning inductively or deductively, looking for patterns in language, and analyzing sayings and expressions commensurate with grade-level learning expectations.

(2) Cross-curricular second language acquisition/listening. The ELL listens to a variety of speakers including teachers, peers, and electronic media to gain an increasing level of comprehension of newly acquired language in all content areas. ELLs may be at the beginning, intermediate, advanced, or advanced high stage of English language acquisition in listening. In order for the ELL to meet grade-level learning expectations across the foundation and enrichment curriculum, all instruction delivered in English must be linguistically accommodated (communicated, sequenced, and scaffolded) commensurate with the student's level of English language proficiency. The student is expected to:

(A) distinguish sounds and intonation patterns of English with increasing ease;

(B) recognize elements of the English sound system in newly acquired vocabulary such as long and short vowels, silent letters, and consonant clusters;

(C) learn new language structures, expressions, and basic and academic vocabulary heard during classroom instruction and interactions;

(D) monitor understanding of spoken language during classroom instruction and interactions and seek clarification as needed;

(E) use visual, contextual, and linguistic support to enhance and confirm understanding of increasingly complex and elaborated spoken language;

(F) listen to and derive meaning from a variety of media such as audio tape, video, DVD, and CD ROM to build and reinforce concept and language attainment;

(G) understand the general meaning, main points, and important details of spoken language ranging from situations in which topics, language, and contexts are familiar to unfamiliar;

(H) understand implicit ideas and information in increasingly complex spoken language commensurate with grade-level learning expectations; and

(I) demonstrate listening comprehension of increasingly complex spoken English by following directions, retelling or summarizing spoken messages, responding to questions and requests, collaborating with peers, and taking notes commensurate with content and grade-level needs.

(3) Cross-curricular second language acquisition/speaking. The ELL speaks in a variety of modes for a variety of purposes with an awareness of different language registers (formal/informal) using vocabulary with increasing fluency and accuracy in language arts and all content areas. ELLs may be at the beginning, intermediate, advanced, or advanced high stage of English language acquisition in speaking. In order for the ELL to meet grade-level learning expectations across the foundation and enrichment curriculum, all instruction delivered in English must be linguistically accommodated (communicated, sequenced, and scaffolded) commensurate with the student's level of English language proficiency. The student is expected to:

(A) practice producing sounds of newly acquired vocabulary such as long and short vowels, silent letters, and consonant clusters to pronounce English words in a manner that is increasingly comprehensible;

(B) expand and internalize initial English vocabulary by learning and using high-frequency English words necessary for identifying and describing people, places, and objects, by retelling simple stories and basic information represented or supported by pictures, and by learning and using routine language needed for classroom communication;

(C) speak using a variety of grammatical structures, sentence lengths, sentence types, and connecting words with increasing accuracy and ease as more English is acquired;

(D) speak using grade-level content area vocabulary in context to internalize new English words and build academic language proficiency;

(E) share information in cooperative learning interactions;

(F) ask and give information ranging from using a very limited bank of high-frequency, high-need, concrete vocabulary, including key words and expressions needed for basic communication in academic and social contexts, to using abstract and content-based vocabulary during extended speaking assignments;

(G) express opinions, ideas, and feelings ranging from communicating single words and short phrases to participating in extended discussions on a variety of social and grade-appropriate academic topics;

(H) narrate, describe, and explain with increasing specificity and detail as more English is acquired;

(I) adapt spoken language appropriately for formal and informal purposes; and

(J) respond orally to information presented in a wide variety of print, electronic, audio, and visual media to build and reinforce concept and language attainment.

(4) Cross-curricular second language acquisition/reading. The ELL reads a variety of texts for a variety of purposes with an increasing level of comprehension in all content areas. ELLs may be at the beginning, intermediate, advanced, or advanced high stage of English language acquisition in reading. In order for the ELL to meet grade-level learning expectations across the foundation and enrichment curriculum, all instruction delivered in English must be linguistically accommodated (communicated, sequenced, and scaffolded) commensurate with the student's level of English language proficiency. For Kindergarten and Grade 1, certain of these student expectations apply to text read aloud for students not yet at the stage of decoding written text. The student is expected to:

(A) learn relationships between sounds and letters of the English language and decode (sound out) words using a combination of skills such as recognizing sound-letter relationships and identifying cognates, affixes, roots, and base words;

(B) recognize directionality of English reading such as left to right and top to bottom;

(C) develop basic sight vocabulary, derive meaning of environmental print, and comprehend English vocabulary and language structures used routinely in written classroom materials;

(D) use prereading supports such as graphic organizers, illustrations, and pretaught topic-related vocabulary and other prereading activities to enhance comprehension of written text;

(E) read linguistically accommodated content area material with a decreasing need for linguistic accommodations as more English is learned;

(F) use visual and contextual support and support from peers and teachers to read grade-appropriate content area text, enhance and confirm understanding, and develop vocabulary, grasp of language structures, and background knowledge needed to comprehend increasingly challenging language;

(G) demonstrate comprehension of increasingly complex English by participating in shared reading, retelling or summarizing material, responding to questions, and taking notes commensurate with content area and grade level needs;

(H) read silently with increasing ease and comprehension for longer periods;

(I) demonstrate English comprehension and expand reading skills by employing basic reading skills such as demonstrating understanding of supporting ideas and details in text and graphic sources, summarizing text, and distinguishing main ideas from details commensurate with content area needs;

(J) demonstrate English comprehension and expand reading skills by employing inferential skills such as predicting, making connections between ideas, drawing inferences and conclusions from text and graphic sources, and finding supporting text evidence commensurate with content area needs; and

(K) demonstrate English comprehension and expand reading skills by employing analytical skills such as evaluating written information and performing critical analyses commensurate with content area and grade-level needs.

(5) Cross-curricular second language acquisition/writing. The ELL writes in a variety of forms with increasing accuracy to effectively address a specific purpose and audience in all content areas. ELLs may be at the beginning, intermediate, advanced, or advanced high stage of English language acquisition in writing. In order for the ELL to meet grade-level learning expectations across foundation

and enrichment curriculum, all instruction delivered in English must be linguistically accommodated (communicated, sequenced, and scaffolded) commensurate with the student's level of English language proficiency. For Kindergarten and Grade 1, certain of these student expectations do not apply until the student has reached the stage of generating original written text using a standard writing system. The student is expected to:

(A) learn relationships between sounds and letters of the English language to represent sounds when writing in English;

(B) write using newly acquired basic vocabulary and content-based grade-level vocabulary;

(C) spell familiar English words with increasing accuracy, and employ English spelling patterns and rules with increasing accuracy as more English is acquired;

(D) edit writing for standard grammar and usage, including subject-verb agreement, pronoun agreement, and appropriate verb tenses commensurate with grade-level expectations as more English is acquired;

(E) employ increasingly complex grammatical structures in content area writing commensurate with grade-level expectations, such as:

(i) using correct verbs, tenses, and pronouns/antecedents;

(ii) using possessive case (apostrophe s) correctly; and

(iii) using negatives and contractions correctly;

(F) write using a variety of grade-appropriate sentence lengths, patterns, and connecting words to combine phrases, clauses, and sentences in increasingly accurate ways as more English is acquired; and

(G) narrate, describe, and explain with increasing specificity and detail to fulfill content area writing needs as more English is acquired.

ELPS Integration into Lesson Planning: Self-Assessment

Rate the current level of integration of the English Language Proficiency standards in your lessons.

A: Always
M: Mostly
S: Sometimes
N: Never

Indicator	A	M	S	N	Comments/Questions
I am aware of my district and school's program goals for ELLs.					
I am aware of specific instructional strategies to support ELLs in attaining English language proficiency.					
Students have opportunities to interact socially in my classroom.					
Students interact using academic English about key concepts in my classroom.					
Students read and write using academic English in my classroom.					
I set language objectives for my students.					
I have integrated the ELPS student expectations into my lessons.					
English learners have opportunities to build vocabulary and concept knowledge.					

How to Read the Cross-Curricular Student Expectations

Cross-curricular student expectations are organized into five categories for second language acquisition:

1. learning strategies
2. listening
3. speaking
4. reading
5. writing

The knowledge and skills statement describes the intentions of the student expectations included in this section.

5) **Cross-curricular second language acquisition/writing.** *The ELL writes in a variety of forms with increasing accuracy to effectively address a specific purpose and audience in all content areas. ELLs may be at the beginning, intermediate, advanced, or advanced high stage of English language acquisition in writing. In order for the ELL to meet grade-level learning expectations across foundation and enrichment curriculum, all instruction delivered in English must be linguistically accommodated (communicated, sequenced, and scaffolded) commensurate with the student's level of English language proficiency*. **For Kindergarten and Grade 1, certain of these student expectations do not apply until the student has reached the stage of generating original written text using a standard writing system.** *The student is expected to:*

(A) learn relationships between sounds and letters of the English language to represent sounds when writing in English;

(B) write using newly acquired basic vocabulary and content-based grade-level vocabulary;

(C) spell familiar English words with increasing accuracy, and employ English spelling patterns and rules with increasing accuracy as more English is acquired;

(D) edit writing for standard grammar and usage, including subject-verb agreement, pronoun agreement, and appropriate verb tenses commensurate with grade-level expectations as more English is acquired

Note that some student expectations do not apply for students at early levels of literacy.

Each student expectation is listed individually by letter. These expectations can be used for creating curriculum frameworks, creating and documenting lesson plans, and writing language objectives for English learners.

ELPS Integration Plan for Teachers

1. Identify language proficiency levels of all ELLs.

2. Identify appropriate linguistic accommodations and strategies for differentiating instruction.

3. Take steps to build a language rich interactive classroom.

4. Identify cross-curricular student expectations of the ELPS (subsection c) that could be integrated as language objectives into existing content area instruction.

5. Create focused lesson plans that target academic language and concept development.

Seven Steps to Building a Language Rich Interactive Classroom

1. Teach students language and strategies to use when they don't know what to say.	1B Monitor language with self-corrective techniques 1D Speak using learning strategies 1F Use accessible language to learn new language 1H Expand repertoire of learning strategies to acquire new language 2D Monitor understanding and seek clarification 2E Use linguistic support to confirm and enhance understanding
2. Encourage students to speak in complete sentences.	1G Distinguish formal and informal English 3A Practice speaking using English sound system in new vocabulary 3C Speak using a variety of sentence structures 3D Speak using grade level vocabulary in context 3F Speak using common and content area vocabulary 3I Use oral language for formal and informal purposes
3. Randomize and rotate who is called on so students of all language levels can participate.	1G Distinguish formal and informal English 3A Practice speaking using English sound system in new vocabulary 3C Speak using a variety of sentence structures 3D Speak using grade level vocabulary in context 3F Speak using common and content area vocabulary 3I Use oral language for formal and informal purposes
4. Use response signals for students to monitor their own comprehension.	1B Monitor language with self-corrective techniques 2D Monitor understanding and seek clarification 2E Use linguistic support to confirm and enhance understanding 2I Demonstrate listening comprehension
5. Use visuals and a focus on vocabulary to build background.	1A Use prior knowledge to learn new language 1C Use techniques to learn new vocabulary 2A Distinguish sound and intonation 2B Recognize English sound system in new vocabulary 2F Derive meaning from a variety of media 3J Respond orally to a variety of media sources 4A Learn relationships of sounds and letters in English 4C Develop sight vocabulary and language structures 5C Spell familiar English words
6. Have students participate in structured reading activities.	4B Recognize directionality of English text 4D Use pre-reading supports 4E Read linguistically accommodated materials 4F Use visual and contextual supports to read text 4G Show comprehension of English text individually and in groups 4H Read silently with comprehension 4I Show comprehension through basic reading skills 4J Show comprehension through inferential skills 4K Show comprehension through analytical skills
7. Have students participate in structured conversation and writing activities.	**Conversation** 1E Use and reuse basic and academic language 2C Learn language heard in interactions and instruction 2H Understand implicit ideas and information 2G Understand general meaning, main points, and details of spoken language 3B Use new vocabulary in stories, descriptions, and classroom communication 3G Orally express opinions ideas and feelings 3E Share in cooperative groups 3H Orally narrate, describe and explain **Writing** 5A Learn relationships between sounds and letters when writing 5B Write using newly acquired vocabulary 5D Edit writing 5E Employ complex grammatical structures 5F Write using variety of sentence structures and words 5G Narrate, describe, and explain in writing

Language Objectives Aligned to Cross-Curricular Student Expectations

(subsection c)

Learning Strategies	
1A: Use what they know about ___ to predict the meaning of ... 1B: Check how well they are able to say ... 1C: Use ___ to learn new vocabulary about... 1D: Use strategies such as ___ to discuss...	1E: Use and reuse the words/phrases ___ in a discussion/writing activity about... 1F: Use the phrase ___ to learn the meaning of ... 1G: Use formal/informal English to describe... 1H: Use strategies such as ___ to learn the meaning of...
Listening	**Speaking**
2A: Recognize correct pronunciation of .... 2B: Recognize sounds used in the words ... 2C: Identify words and phrases heard in a discussion about ... 2D: Check for understanding by ..../Seek help by ... 2E: Use supports such as ___ to enhance understanding of ... 2F: Use __ (media source) to learn/review .... 2G: Describe general meaning, main points, and details heard in ... 2H: Identify implicit ideas and information heard in ... 2I: Demonstrate listening comprehension by...	3A: Pronounce the words ___ correctly. 3B: Use new vocabulary about ___ in stories, pictures, descriptions, and/or classroom communication ... 3C: Speak using a variety of types of sentence stems about ... 3D: Speak using the words___ about... 3E: Share in cooperative groups about ... 3F: Ask and give information using the words... 3G: Express opinions, ideas and feelings about ___ using the words/phrases... 3H: Narrate, describe and explain .... 3I: Use formal/informal English to say ... 3J: Respond orally to information from a variety of media sources about...
Reading	**Writing**
4A: Identify relationships between sounds and letters by... 4B: Recognize directionality of English text. 4C: Recognize the words/phrases.... 4D: Use prereading supports such as___ to understand ... 4E: Read materials about ___ with support of simplified text/visuals/word banks as needed. 4F: Use visual and contextual supports to read ... 4G: Show comprehension of English text about ... 4H: Demonstrate comprehension of text read silently by... 4I: Show comprehension of text about ___ through basic reading skills such as ... 4J: Show comprehension of text/graphic sources about ___ through inferential skills such as ... 4K: Show comprehension of text about ___ through analytical skills such as ...	5A: Learn relationships between sounds and letters when writing about ... 5B: Write using newly acquired vocabulary about... 5C: Spell English words such as ... 5D: Edit writing about ... 5E: Use simple and complex sentences to write about ... 5F: Write using a variety of sentence frames and selected vocabulary about ... 5G: Narrate, describe, and explain in writing about ...

Examples of Content and Language Objectives Aligned to the Math TEKS and Cross-Curricular Student Expectations

(subsection c)

TEKS	Content Objective	ELPS	Language Objective
	Elementary		
K.9C	I can identify circles, triangles, rectangles and squares.	1C	I can use Vocabulary Alive to learn new vocabulary about shapes.
1.1B	We will use base ten blocks to order whole numbers.	2C	We will identify sounds in the words ones and tens heard in a discussion about place value.
2.11A	The student will construct picture graphs.	3A	The student will pronounce the words, collect data, graph and picture graph correctly.
3.7B	I can identify and describe patterns in a table.	3E	I can share in a cooperative group what I know about patterns.
4.12B	I will solve elapsed time problems by using a stop watch.	4F	I will use visual and contextual supports to read problems about elapsed time.
5.3E	The learner will add fractions with common denominators by using fraction bars.	5G	The learner will explain in writing how to add fractions with common denominators by using the words fraction bar, common denominator, and add.
	Middle School		
6.4B	We will generate perimeter and area formulas of rectangles by using tables of data.	1G	We will use formal/informal English to describe the difference between perimeter and area.
7.7A	The student will use ordered pairs to locate and name points on a coordinate plane.	2I	The student will demonstrate listening comprehension by playing a game of Coordinate Geometry Bingo.
7.11A	I will select among a line graph, bar graph and circle graph to display collected data.	3G	I will express opinions about which graph to select by using the sentence stem, "I think the best choice is _____ because…"
8.4A	The student will be able to generate 3 different representations (table, graph, equation) of a given problem situation.	4G	The student will show comprehension of English text by reading a problem and summarizing its meaning in 3 different representations.
8.7C	The learner will derive the Pythagorean Theorem by using centimeter grid paper.	5B	The learner will write using newly acquired vocabulary about right triangles.
	High School		
A.2B	We will determine reasonable domains and ranges of given situations.	1E	We will use and reuse the words domain, range, in continuous and discrete situations.
A.6D	The student will write equations of lines given a point and a slope, two points, or a slope and y-intercept.	2H	The student will identify implicit ideas and information heard in discussion by participating in a Carousel Sharing Activity.
G.9D	I will analyze the characteristics of polyhedra and other 3D figures by exploring concrete models.	3H	I will describe the characteristics of polyhedra and other 3D figures by using the stem, "A characteristic is…"
AII.5B	The learner will sketch graphs of conic sections.	ID	The learner will use strategies such and comparing and contrasting to discuss similarities and differences between graphs of conic sections.
MMA.6B	The student will be able to investigate home financing by using amortization models.	4E	The student will be able to read materials about home financing with support of simplified text.
P.3A	We will investigate trigonometric functions.	5B	We will write using newly acquired vocabulary about trigonometric functions.

ELPS Lesson Plan Template

Grade: **Topic:**
Subject: **Date:**

Content Objective *(Aligned with TEKS)*:	Language Objective *(Aligned with ELPS)*:
Vocabulary:	Visuals, Materials & Texts:

Activities	**Review & Checks for Understanding:**
Activating Prior Knowledge *(Processes, Stems and Strategies)*: Building Vocabulary and Concept Knowledge *(Processes, Stems and Strategies)*: Structured Conversation and Writing *(Processes, Stems and Strategies)*:	*(Response Signals, Writing, Self-Assessment Student Products, etc,)*

ELPS Lesson Plan Sample (Elementary)

Grade: 1st **Topic:** Measuring with Nonstandard Units

Subject: Math **Date:** 7/19/2009

Content Objective ***(Aligned with TEKS)*:** **(7A)** I will estimate and measure the length of a pencil, my foot and desk by using color tiles.	**Language Objective** ***(Aligned with ELPS)*:** **(1H)** I can use Vocabulary Alive to learn the meaning of measure, estimate, length, and exact.
Vocabulary: measure, estimate, length, exact	**Visuals, Materials & Texts:** color tiles, chart paper, markers, response boards, dry erase markers, vocabulary PPT

Activities	**Review & Check for Understanding:** ***(Response Signals, Writing, Student Product, Student Self-assessment.)***
Activating Prior Knowledge ***(Processes, Stems, Strategies)*:** Teacher Questions ▪ What does measure mean to you? ▪ What are objects you can measure in the room? ▪ What can you use to measure objects in the room? Whip Around Expressing prior knowledge stems : *Measure means...* *I can measure...* *I can use a _____ to measure a _____.* **Building Vocabulary and Concept Knowledge** ***(Processes, Stems, Strategies)*:** ▪ Use vocabulary PowerPoint to introduce the lesson's key vocabulary ▪ Students create a four corners vocabulary chart for each term ▪ Students participate in vocabulary alive ▪ Model examples where color tiles are used to estimate and measure the length of objects in the room ▪ Students use color tiles to estimate and measure the length of a pencil, their foot, and their desk Descriptive language stems: *The word _____ means...* *An example of the word _____ is...* *The picture I drew for the word _____ is...* *I think _____ measures _____ color tiles in length.* *The/My _____ measures exactly_____ color tiles in length.* **Structured Conversation and Writing** ***(Processes, Stems, Strategies)*:** Write, Inside/Outside Circles Persuasive Language Stems: *My estimate is correct because...* *I do/do not think her guess is correct because...* *I do/do not think ____ measures ____ because...* *I estimate the _____ to be _____ color tiles long because...*	Orally review previous lesson Agree/Disagree Listen to student conversations Provide sufficient wait time for students to formulate responses Observe student work Do students know what measure means or is there a need to build background? Are students' estimates reasonable? Are students using the lesson's vocabulary to express their thoughts? Review key content by showing students objects and having them estimate the length by writing the answer on response boards

ELPS Lesson Plan Sample (Middle School)

Grade: 6th **Topic:** Circle Graphs

Subject: Math **Date:** July 31, 2009

Content Objective ***(Aligned with TEKS):*** **(10C)** I will sketch circle graphs to display data.	**Language Objective** ***(Aligned with ELPS):*** **()** I will write using newly acquired vocabulary about bar and circle graphs.
Vocabulary: graph, bar graph, circle graph, data, percents, represent	**Visuals, Materials & Texts:** graphing chart paper, markers, chart paper, calculator (opt.)

Activities	Review & Check for Understanding: *(Response Signals, Writing, Student Product, Student Self-assessment.)*
Activating Prior Knowledge ***(Processes, Stems, Strategies):*** Teacher Questions ▪ What do you know about graphs? Graffiti Write Expressing prior knowledge stems : *A graph is...* *One thing I remember about graphs is...* *An example of a graph is...* *A type of graph is...* **Building Vocabulary and Concept Knowledge** ***(Processes, Stems, Strategies):*** ▪ Introduce the lesson's vocabulary by following the first three steps of the Six Step Vocabulary Process. ▪ Collect data by having students complete a survey: Favorite Soft Drink ▪ With the collected data, students create a human bar and circle graph. Have students draw a sketch of both the bar graph and circle graph. ▪ Have students complete another survey: Favorite Sport ▪ Students work in groups of 3 to create a bar graph of the collected data ▪ Students create a circle graph with the bar graph. ▪ In groups, students complete a Venn diagram to determine the differences and similarities between bar and circle graphs. Descriptive language stems: *_____ means...* *A picture that represents _____ is...* *An example of a bar graph is ...* *My favorite soft drink is...* *A difference between a bar and circle graph is...* *A similarity between a bar and circle graph is...* **Structured Conversation and Writing** ***(Processes, Stems, Strategies):*** Write, Think Pair Share Persuasive Language Stems: *I think the teacher's favorite drink is ____ because...* *It is best to represent data on a circle graph when...*	Review previous lesson's vocabulary by participating in a game of Act It Out. Listen to student conversations Observe student work Use Number Wheels to determine student's favorite drink/sport and for them to vote which is teacher's favorite drink. Can students connect percents with circle graphs? Can students describe the similarities and differences between circle and bar graphs? Outcome Sentences *I learned...* *I liked...* *I wonder...* *I think...*

ELPS Lesson Plan Sample (High School)

Grade: 9th **Topic:** Linear Functions and Math Representations

Subject: Math **Date:** 11/15/08

Content Objective ***(Aligned with TEKS)*****: (A.5A)**	**Language Objective** ***(Aligned with ELPS)*****: (3E)**
SWBAT determine whether or not the pool pattern can be represented by a linear function.	SWBAT orally defend whether the pool pattern is discrete or continuous by participating in a Think, Write, Pair, Share activity.
Vocabulary: linear function, domain, range, independent variable, dependent variable, slope, y-intercept, continuous situation, discrete, math representations	**Visuals, Materials & Texts:** Student created four corner vocabulary, color tiles, overhead color tiles, math representations graphic organizer, key vocabulary cards with nonlinguistic representations, chart paper, markers, index cards

Activities	**Review & Check for Understanding:** ***(Response Signals, Writing, Student Product, Student Self-assessment.)***
Activating Prior Knowledge ***(Processes, Stems, Strategies)*****:** Think, Pair, Share Expressing prior knowledge stems : *The five mathematical representations are_____, _____, _____, _____, and _____.* *Mathematical representations are useful for_____.*	Orally review previous lesson by having students participate in a Carousel Sharing activity
Building Vocabulary and Concept Knowledge ***(Processes, Stems, Strategies)*****:** ▪ Students create concept maps in groups for the terms: linear function, slope, y-intercept, domain, range, math representations ▪ Create pool patterns with color tiles ▪ Use MR graphic organizer to represent pool patterns in pictorial, written, tabular, graphical, and symbolic forms Descriptive language stems: *The main idea of my concept map is...* *The difference between a table and a graph is...* *The attributes of linear functions are...* *The rate of change is____ and represents _____...* *The y-intercept is ____ and represents _____...*	Listen to student conversations Observing student work Check for titles, labels and intervals on graphs Can students recognize the function rule in the concrete or pictorial representations? Do students complete all sections of MR graphic organizer?
Structured Conversation and Writing ***(Processes, Stems, Strategies):*** Write, Think, Pair, Share, Response Boards Persuasive Language Stems: *"I do/do not think the ordered pairs in the graph should be connected because...* *The pool pattern is/is not a linear function because...* *This problem situation is discrete/continuous because...*	Thumbs: Up/Down/In the Middle

Representing Math in Multiple Ways

Concrete/Picture	Equation/Function Rule

Verbal & Written Description

Table	Graph

	Process Column	

Created by Araceli Avila. Used with permission.

ELPS Lesson Plan Activity Guide

(subsection c)

Instructional Strategies	ELPS Student Expectation Summaries	Classroom Strategies/Techniques	
Activating Prior Knowledge	1A Use prior knowledge to learn new language 1F Use accessible language to learn new language 4D Use pre-reading supports 4E Read linguistically content area accommodated materials 4F Use visual, contextual, and peer supports to read text	• Anticipation Guides • Advance Organizers • Backwards Bookwalk • Chunking Input • Graffiti Write • Graphic Organizers	• KWL • Math Sort • Prediction Café • Scanning • Vis. Literacy Frames • Visuals/Video
Building Vocabulary and Concept Knowledge	1E Use and reuse basic and academic language 1C Use techniques to learn new vocabulary 1H Expand repertoire of learning strategies to acquire language 2B Recognize English sound system in new vocabulary 2F Derive meaning from a variety of media 3A Practice speaking using English sound system in new vocabulary 3B Use new vocabulary in oral communication 4A Learn relationships of sounds and letters in English 4B Recognize directionality of English text 4G Show comprehension of English text individually and in groups 4H Read silently with comprehension 4I Show comprehension through basic reading skills 4J Show comprehension through inferential skills 4K Show comprehension through analytical skills 4C Develop sight vocabulary and language structures	• Cloze Sentences • Concept Attainment • Comprehension Strategies • Creating Words • Dirty Laundry • DRTA • Expert/Novice • Guess Your Corner • Hi-lo readers • Homophone/ • Homograph Sort • List/Sort/Label • Mix and Match	• Nonlinguistic Rep. • Prefixes, Suffixes and Roots • QtA • QAR • SQP2RS • Self-Assessment of Word Knowledge • Word Analysis • Think Alouds • Word Generation • Word Sorts • Word Walls • Vocabulary Alive
Structured Conversation	1B Monitor language with self-corrective techniques 1G Distinguish between formal and informal English 2D Monitor understanding and seek clarification 2E Use support to confirm and enhance understanding 1D Speak using learning strategies 2C Learn language heard in interactions and instruction 2I Demonstrate listening comprehension 2A Distinguish sound and intonation 2G Understand general meaning, main points, and details 2H Understand implicit ideas and information 3C Speak using a variety of sentence structures 3D Speak using grade level content area vocabulary in context 3E Share in cooperative groups 3F Ask and give information using common and content area vocabulary 3G Orally Express opinions ideas and feelings 3H Orally Narrate, describe and explain 3I Use oral language for formal and informal purposes 3J Respond orally to a variety of media sources	• Accountable Conversation Stems • Instr. Conversation • Literature Circles • Num. Heads Together • Perspective-Based Activities • Question Answer Relationship (QAR)	• QSSSA • Response Triads • Reciprocal Teaching • Structured Conv. • Structured Academic Controversy • Think, Pair, Share, • Tiered Resp. Stems • W.I.T.
Writing	5A Learn relationships between sounds and letters when writing 5B Write using basic and content area vocabulary 5C Spell familiar English words accurately 5D Edit writing for standard grammar and usage 5E Employ complex grammatical structures in content area 5F Write using variety of sentence structures and words 5G Narrate, describe, and explain in writing	• Book Reviews • Contextualized Grammar Instruction • Chat Room • Daily Oral Language • Double Entry Journals • Genre Analysis and Imitation	• Learning Logs • RAFT • Roundtable • Sentence Stems • Sentence Mark Up • Sentence Sorts • Summary Frames • Unit Study for ELLs

Comparing Whole Numbers in a Conga Line

Math TEKS						ELPS					
☒	Nm & Op	$1^{st} - 5^{th}$	☐	Algebra		☒	Lng Str	1E	☒	Listen	2I
☐	Geometry		☐	Measure		☒	Speak	3BCD	☐	Read	
☐	Pro & Sts		☒	Pro & Tls	$1^{st} - 5^{th}$	☐	Write				

Content Objective(s)	Language Objective(s)
I can use place value to compare whole numbers.	I can demonstrate listening comprehension by participating in a conga line activity.

Key Vocabulary

Content	Process/Functional
• whole numbers • place value • ones place • tens place • hundreds place • thousands place • greater than • less than	• compare • order • base ten blocks

Supplementary Materials

- place value mat
- base ten blocks
- whole number cards

Activity Instructions

Preparation

- Copy the whole number cards on card stock (3-4 copies of each page).
- Cut the cards and place them a gallon size plastic bag.

Process

1. In order to determine what students know about the word *compare*, activate students' prior knowledge by asking them "Who is the tallest person in your house?" Provide them with sufficient time to formulate a response and have them share their reflection with a partner. Once all pairs are done sharing, randomly select several students to share with the whole class.
2. Inform student in order for them to determine who the tallest person at their house was they had to compare the heights of the people who live with them. If students do not have prior knowledge of what compare means, build their background by having them compare their heights by lining up in front of the class from tallest to shortest.
3. Introduce the lesson's content and language objectives.
4. Use the vocabulary PPT to introduce/review the lesson's key vocabulary.
5. Depending on the grade level, select 2-4 number cards from the plastic bag. Tell students you are trying to create the largest number with the cards. Use base ten blocks and a place value mat and model for them how to form the largest number. (Repeat this step as many times as necessary.)
6. Provide each student with 2-4 whole number cards, base ten blocks and a place value mat. Instruct students to form the largest number possible with their cards. Once all students are done, ask them to turn to their partner and say "My whole number is …" Pairs need to determine who has the largest number.
7. Have students trade one card with their partner and repeat step 6.
8. Tell students they are going to play a game called "Who has the greatest number?" by participating in a conga line activity. Divide students into two groups and have them form two lines facing one another. Give each child 2-4 cards.
9. Tell students the objective of the game is to create the largest number with the cards and compare the number with their partner to determine who has the greatest number. Student from Group 1 will speak first and say, "My number is…" Student from Group 2 will listen and then say "My number is…" Student with the largest number wins by saying, "I have the greatest number."
10. Have the student at the farthest end of Group 1 run down the middle of the two lines and join the other end of Group 1. Have all students in Group 1 move one person in. Group 2 is stationary. Repeat step 9 at least 4-5 times.
11. Review key vocabulary and content and language objectives.

Whole Number Cards

1	1	1	1
2	2	2	2
3	3	3	3
4	4	4	4
5	5	5	5

Whole Number Cards

6	6	6	6
7	7	7	7
8	8	8	8
9	9	9	9

Multiple Representations Card Game

Math TEKS						ELPS					
☐	Nm & Op		☒	Algebra	6.3B	☒	Lng Str	1E	☒	Listen	2CG
☐	Geometry		☐	Measure		☒	Speak	3BC**H**	☒	Read	4GHI
☐	Pro & Sts		☒	Pro & Tls	6.12A	☒	Write	5F			

Content Objective(s)	Language Objective(s)
We will represent ratios in multiple ways.	We will explain why we won a Ratio Card Game by stating: • I win because the ratio _____ is the same as _____, _____ and _____.

Key Vocabulary

Content	Process/Functional
• ratio • ratio in fraction notation • ratio in odds notation • ratio in word notation	• explain • represent • representation

Supplementary Materials

- vocabulary PPT
- ratio cards
- ratio lab sheet

Activity Instructions

NOTE: This activity is a variation of the Spoons Card Game.

Preparation

- Copy the ratio cards on card stock paper (5 sets = 20 cards; up to five students can play with 1 set of cards).
- Cut all 20 cards and place them in a quart size plastic bag.
- Copy one ratio lab sheet per student.

Process

1. Activate students' prior knowledge by having students write down three things they know or remember about ratios. Have students share their responses with a partner. Finally select a couple of students to share their responses with the whole class.
2. Introduce the lesson's content and language objectives.
3. Use the vocabulary PPT to review the meaning of ratio and the different ways a ratio can be represented.
4. Group students in sets of 3 – 5 and tell them they are going to play a ratio card game in which the objective is to be the first player to collect the four cards that represent the same ratio.
5. Provide each group a set of ratio cards.
6. Assign one group member to be the card dealer.
7. The dealer shuffles cards and distributes four cards to each player.
8. Players look at their cards and determine whether they have four of a kind. If nobody has four of a kind, all players simultaneously select 1 unwanted card and pass card to their left-hand neighbor. The player who passes the card to the dealer needs to place the unwanted card under the pile of leftover cards. The dealer picks up a card from the top of the pile.
9. The process continues until someone gets four of a kind. In order to be declared the winner of the round, the person with four of a kind shouts "Ratio", shows his/her cards to the rest of the group and explains why he/she won. Player uses the sentence frame: "I win because the ratio _____ is the same as _____, _____ and _____."
10. After a player wins the round, all players complete the ratio lab sheet by recording the winner's four of a kind cards.
11. Students continue playing by repeating steps 7 – 10.
12. Review key vocabulary and content and language objectives.

Ratio Lab Sheet

Name of Winner	What was the word problem about?	Fractional Notation	Word Notation	Odds Notation
	The word problem was about…			
	The word problem was about…			
	The word problem was about…			
	The word problem was about…			
	The word problem was about…			
	The word problem was about…			
	The word problem was about…			
	The word problem was about…			
	The word problem was about…			

Ratio Cards

7,349 out of 365	$\frac{7,349}{365}$
7,349:365	 A baby uses an average of seven thousand three hundred forty nine disposable diapers in three hundred sixty five days. What ratio represents this situation?

Ratio Cards

4 to 2	$\frac{4}{2}$
4:2	 Daniela is trying to make rice for dinner. The directions say to use 4 cups of water for every 2 cups of rice. What ratio represents this situation?

Ratio Cards

100 to 2	$\frac{100}{2}$
100:2	 A hummingbird flaps its wings about one hundred times in two seconds. What ratio represents this situation?

Ratio Cards

18 to 15	$\frac{18}{15}$
18:15	 The average human heart beats 18 times every 15 seconds. What ratio represented this situation?

Order It Up! Math Puzzle (Created by Amy Serda-King)

Math TEKS						ELPS					
☒	Nm & Op	7.2E	☐	Algebra		☐	Lng Str	1E	☒	Listen	2I
☐	Geometry		☐	Measure		☒	Speak	3E**H**	☐	Read	
☐	Pro & Sts		☒	Pro & Tls	7.13C	☐	Write				

Content Objective(s)	Language Objective(s)
Students will determine equivalent numerical expressions involving order of operations and exponents when given the answer to the expression.	Students will share with a partner the process of solving a numerical expression by saying: • The equivalent expression is…, • First you solve… • Then you solve… • Finally the result is…

Key Vocabulary		Supplementary Materials	
Content	Process/Functional	• Vocabulary PPT • Order It UP Answer Sheet • Order It Up Cards • Order It Up lab sheet	• Blank paper • Calculators (optional)
• Exponent • Expression • Order of operations			

Activity Instructions

Preparation:

- Copy and cut Order It Up! Cards.
- Place each numerical expression set of cards in an envelope. Make sure you have the actual number written in front of the envelope.
- Copy one Order It Up! Lab sheet per student.

Process:

1. Use the vocabulary PPT to review key vocabulary with students.
2. Review the steps to simplifying expressions using order of operations.
3. Inform students they will play a game where they are given a number and a set of cards that create an equivalent expression for the given number. The goal of the game is for students to manipulate the parts of the expression which equates to the given number.
4. Model the game with the whole class. Provide students the answer to the numerical expression and the parts of the expression on the board.

For example the answer is 3. The parts of the expression are 2^2, 5,+, (, 12,), -18, +,+
The goal is for students to reorganize the parts of the expression so that they form the equivalent numerical expression which provides a result of 3.

$$2^2 + (-18+12) + 5$$

5. Group students in pairs.
6. Pass out a problem to each pair.
7. Pass out a blank sheet of paper so students can use it as scratch paper.
8. Students will **Think** and **Solve** the problem **individually** before discussing it with a partner.
9. Students will utilize the problem solving strategy of Guess and Check to determine the given result. The student who creates the expression first will orally explain the steps using the following sentence frames to their partner. **"The equivalent expression is …", "First you solve….Then you solve…Finally the result is".** Place the sentence frames on the board where students can see them.
10. Students will complete the Order It Up! student lab sheet.
11. Partners will share the process of solving the numerical expression and initial their partner's work to ensure they reviewed the problem using the sentence frames.
12. Continue this process for five more problems.
13. Students will create their own mathematical puzzle expression and challenge their partner to solve the equivalent expression. Students will complete the sentences on the lab sheet.

Order It Up! Lab Sheet

Answer	Numerical Expression	Partner's Initials

A. Create your own mathematical puzzle below.
B. Switch your puzzle with a partner and try to solve the numerical expression.
B. Complete the sentences below.

Result								

The numerical expression is ______________________________.

The result is ____________.

©Helping Math Teachers Implement the ELPS. Used with Permission

Order It Up! Answer Sheet

Beginner	
Page	Equation
1.	$16 = 2\ (5+3)$
2.	$3 = 2 \cdot 4 - 35 \div 7$
3.	$8 = 2 \cdot 3 + 14 \div 7$
4.	$26 = 50 - (15 + 9)$
5.	$1 = (11 - 2\ ) \div 9$
Intermediate	
6.	$12 = (15 \div 3) + 7$
7.	$22 = 3 + 4 + 45 \div 3$
8.	$18 = 6\ (9 - 6)$
9.	$-70 = -7\ (6 + 4)$
10.	$67 = 6^2 + 7 + 3 \cdot 8$
Advanced	
11.	$27 = 5 \cdot 2 - 32 + 7^2$
12.	$-8 = -2\ (36) \div 9$
13.	$17 = 5\ (6 \div 3) + 6 + 1$
14.	$4 = (17 + 3) \div (4 + 1)$
15.	$3 = 3\ (4 \cdot 9) \div 6^2$

Order It Up! Cards

16			
=	2	(	)
5	3	+	

3			
35	•	÷	7
–	4	=	2

Order It Up! Cards

8			
2	=	•	14
+	7	3	÷

26			
(	+	50	9
15	−	)	=

Order It Up! Cards

1			
(	11	−	÷
9	=	2	)

12			
3	)	+	7
÷	15	(	=

Order It Up! Cards

22			
3	+	45	÷
+	3	4	=

18			
=	)	9	6
6	−	(	

Order It Up! Cards

-70			
+	)	=	6
4	-7	(	

67			
+	8	$\bullet$	6^2
7	+	=	3

Order It Up! Cards

27			
2	=	•	7^2
−	5	32	+

-8			
)	-2	36	÷
=	9	(	

Order It Up! Cards

17			
=	5	6	(
÷	)	3	+
+	6	1	

Order It Up! Cards

4			
÷	)	=	17
)	3	+	(
4	(	1	+

Order It Up! Cards

3			
(	3	4	=
9	)	$\div$	6^2
$\bullet$			

Parallel and Perpendicular Lines Sort

Math TEKS						ELPS					
☐	Algebra I		☐	Algebra II		☒	Lng Str	1CF	☐	Listen	2EFG
☒	Geometry	G.7B	☐	Math Md		☒	Speak	3EGJ	☐	Read	
						☐	Write				

Content Objective(s)	Language Objective(s)
Students will use slopes and equations of lines to investigate geometric relationships of parallel and perpendicular lines.	Students will compare and contrast the slopes of linear equations by participating in a math sort activity.

Key Vocabulary

Content	Process/Functional
• slopes • parallel Lines • perpendicular Lines • negative Reciprocals	• determine • classify

Supplementary Materials

- vocabulary PPT
- parallel and perpendicular lines slide show
- parallel and perpendicular lines sort cards and chart
- chart paper
- markers

Activity Instructions

Preparation

- Copy the parallel and perpendicular lines sort cards and chart on card stock. (1 per group)
- Cut the cards and place them in a quart size plastic bag.

Process

1. Activate students' prior knowledge by showing them the parallel and perpendicular lines slide show. Once they have finished watching the slide show, ask them to guess what the slide show is about. Have students shout out their answers. Once it is established the slide show is about parallel and perpendicular lines, place them in groups of 3-4. Provide each group chart paper and markers and have them brainstorm at least 5 characteristics of parallel and perpendicular lines.
2. Introduce lesson's content and language objectives.
3. Use the vocabulary PPT to review the meaning of parallel and perpendicular lines.
4. Model 2 to 4 examples of how to determine if a system is perpendicular, parallel or neither.
5. Have students practice classifying parallel and perpendicular lines. Group students in sets of 3 and tell them they are going to determine if systems of linear equations are perpendicular, parallel or neither by participating in a sorting activity.
6. Provide each group a set of the cards, chart and graphing calculators and have them determine the category of each system of linear equations.
7. Once all groups are finished sorting have groups compare their answers with another group. If answers are incorrect, group members need to defend why the answer is correct or incorrect.
8. Review key vocabulary and content and language objectives.

Parallel and Perpendicular Lines Sort Cards

Parallel Lines	Perpendicular Lines	Neither
$y = \frac{1}{2}x - 4$ $y = \frac{1}{2}x + 2$	$y = \frac{1}{2}x - 4$ $y = -2x + 2$	$y = \frac{1}{2}x - 4$ $y = 2x + 2$
$y = \frac{4}{3}x - 1$ $3y = 4x - 14$	$y = \frac{4}{3}x - 1$ $y = \frac{-3}{4}x - 14$	$y = \frac{-3}{4}x - 1$ $y = \frac{3}{4}x - 14$
$2y = 4x - 12$ $y = 2x + 15$	$2x + y = 6$ $y = \frac{1}{2}x - 1$	$2y = 4x - 12$ $y = -2x + 15$
$y = 5x - 2$ $y = 5x + 1$	$7x + 2y = -9$ $-2x + 7y = 14$	$y = 4x - 12$ $y = -4x + 15$
$3x + 8y = -24$ $8y = -3x + 15$	$3x + 8y = -24$ $-8x + 3y = 15$	$2y = 6x - 12$ $y = 5x + 15$

Parallel and Perpendicular Lines Sort Chart

Parallel Lines	Perpendicular Lines	Neither

Making Inferences from Graphs

Math TEKS						ELPS					
☒	Algebra I	A.2BCD	☐	Algebra II		☒	Lng Str	1AEF	☒	Listen	2DEFGH
☐	Geometry		☐	Math Md		☒	Speak	3CEGJ	☒	Read	4EFGJ
						☒	Write	5G			

Content Objective(s)	Language Objective(s)
Students will interpret situations in terms of given graphs.	Students will infer and make critical judgments from a graph by stating orally and in writing: • I can infer that… • A possible solution is…

Key Vocabulary		Supplementary Materials	
Content	Process/Functional		
• Positive Correlation • Negative Correlation • No Correlation	• Infer • Critical Judgment	• KWL Chart • 4 Corners Vocab Chart • Vocabulary PPT • Beverage Bottle Wasting and Recycling Graph and Students Notes	• Municipal Solid Waste (MSW) in the US—Student Lab Sheet • Waste Situations 1-4 • Index Cards • Chart Paper • Markers

Activity Instructions

Preparation

- Copy and cut the Waste Situations 1-4 and post them around the room (read step 8).
- Copy one KWL, 4 Corners Chart, Student Notes and Student Lab Sheet per student.

Process

1. Provide each student with a KWL Chart. Write the words Positive Correlation and Negative Correlation on the board. Provide students 2 minutes to write or draw under the K section what they know about these two words. Select a couple of students to share with the class what they wrote.
2. Use the vocabulary PPT to introduce the lesson's key terms. Have students process the meaning of the terms by completing the 4 Corner Vocabulary Chart.
3. Introduce content and language objectives.
4. To set the stage for today's main activity, play the first 5-7 minutes of the video, June 28 Landfills and Envelopes, found at: http://www.truveo.com/June-28-Landfills-and-Envelopes/id/288230384801046720. Facilitate a discussion about the video.
5. Distribute and display the Bottled Beverage Wasting and Recycling Graph and Student Notes Sheet. Facilitate a discussion by asking students the facilitation questions and telling them to take notes. **Facilitation Questions:** What is this graph about? What can you infer from the graph? Is the situation demonstrating any type of correlation? How do you know? What are the situation's variables? What is the domain and range? Is the situation continuous or discrete? Based on the graph's data, is this situation a concern to our environment? Why or Why not?
6. Provide students an opportunity to examine other waste problem situations by participating in a Carousel Activity. Around the room, post 8 pieces of chart paper. Number the pieces of chart paper 1-8. On Chart Papers 1 and 5, glue Situation 1: Metal Wasting and Recycling; Chart Papers 2 and 6, glue Situation 2: Plastic Wasting and Recycling; Chart Papers 3 and 7, glue Situation 3: Paper and Paperboard Wasting and Recycling; and Chart Papers 4 and 8, glue Situation 4: Glass Wasting and Recycling.
7. Ask students to count off 1-8. Based on their assigned number, have them go to their corresponding numbered chart paper. Ask each group to study their chart paper's graph and write one thing they infer from the graph's data. Have groups rotate counterclockwise, study the new graph and write one thing they infer about the graph. Repeat this process two more times.
8. After the third rotation, have groups rotate one more time and provide them with the Municipal Solid Waste (MSW) in the US—Student Lab Sheet. Have them answer the questions on the Lab Sheet.
9. Have a person from each group share their group's solution to the problem situation with the whole group.
10. Provide students an index card and ask them to identify the problem situation that merits the most attention. Once they identify the most important problem situation, have them write the name of the situation on the index card and explain why they think this situation merits the most attention. Sentence Stem: The situation that merits the most attention is __________ because …
11. Review lesson's vocabulary and close lesson by reviewing content and language objectives.

Beverage Bottle Wasting and Recycling Graph and Student Notes

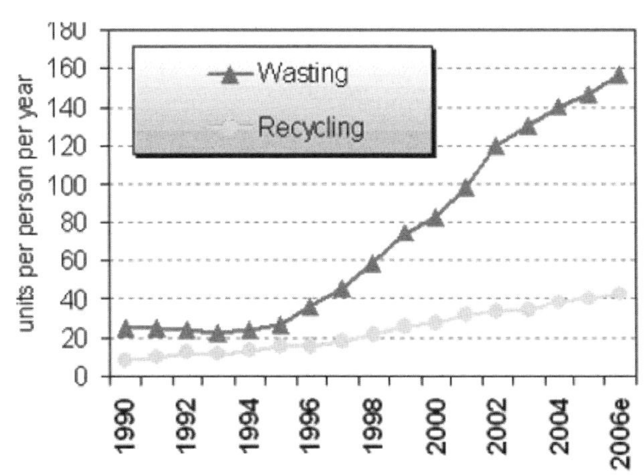

http://www.sustainablebusiness.com/imageupload/Plastics-recycling-graph.gif

What is this graph about?	***What can you infer about this graph?***	***Is there a correlation between the millions of tons of waste and the year number?***
This graph is about…	I can infer that…	The situation has a ____________ correlation.
What are the situation's variables?	***What is the domain and range***	***Is the graph's data a concern for our environment?***
The variables are…	The domain is… The range is… ***Is the situation discrete or continuous?*** The situation is…	The data on the graph is/is not a concern for our environment because…

Municipal Solid Waste (MSW) in the US—Student Lab Sheet

What is this graph about? This graph is about…	***What can you infer about this graph?*** I can infer that…	***Is there a correlation between the millions of tons of waste and the year number?*** The situation has a ___________ correlation.
What are the situation's variables? The variables are…	***What is the domain and range*** The domain is… The range is… ***Is the situation discrete or continuous?*** The situation is…	***Is the graph's data a concern for our environment?*** The data on the graph is/is not a concern for our environment because…
A possible solution to this situation is…		

Situation 1: Metal Wasting and Recycling

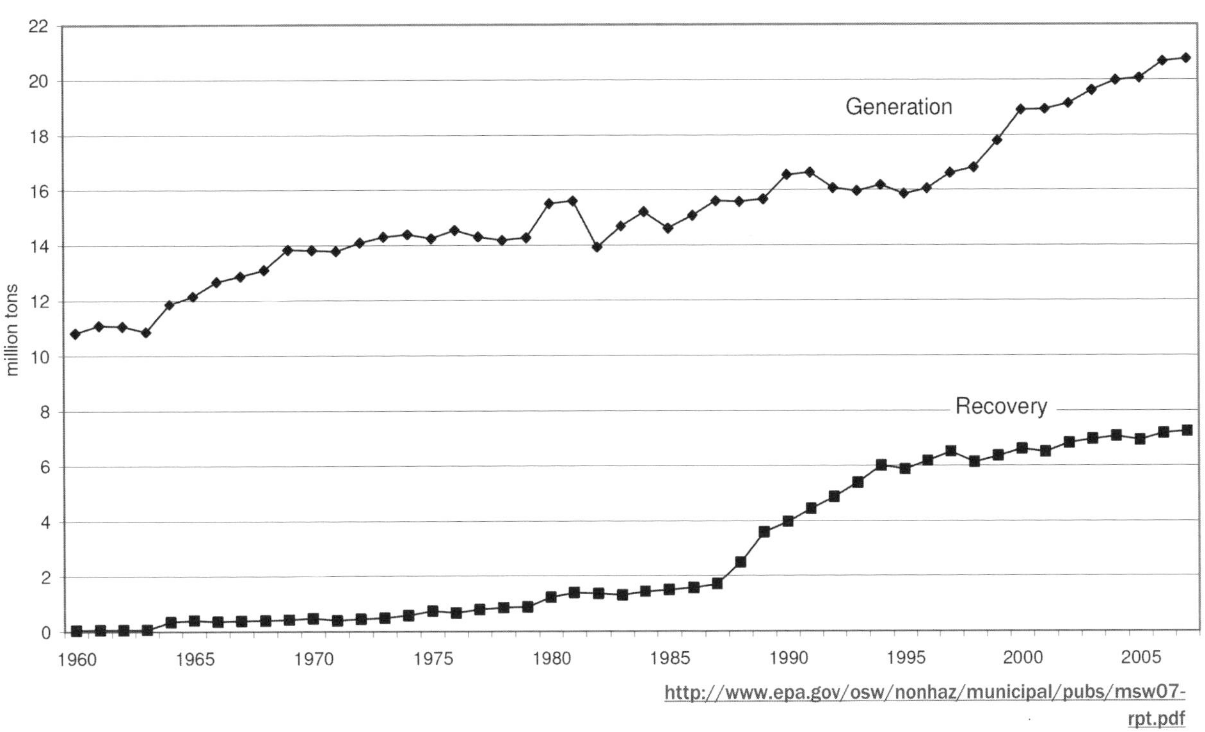

http://www.epa.gov/osw/nonhaz/municipal/pubs/msw07-rpt.pdf

Situation 2: Plastic Wasting and Recycling

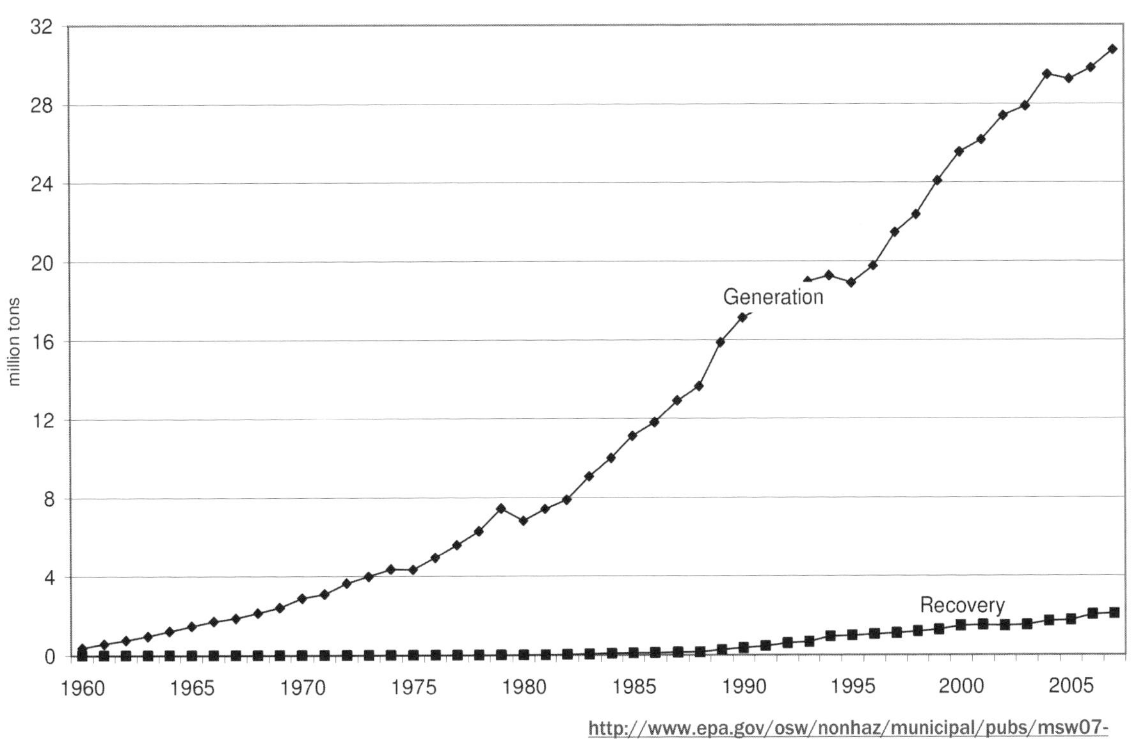

http://www.epa.gov/osw/nonhaz/municipal/pubs/msw07-rpt.pdf

Situation 3: Paper and Paperboard Wasting and Recycling

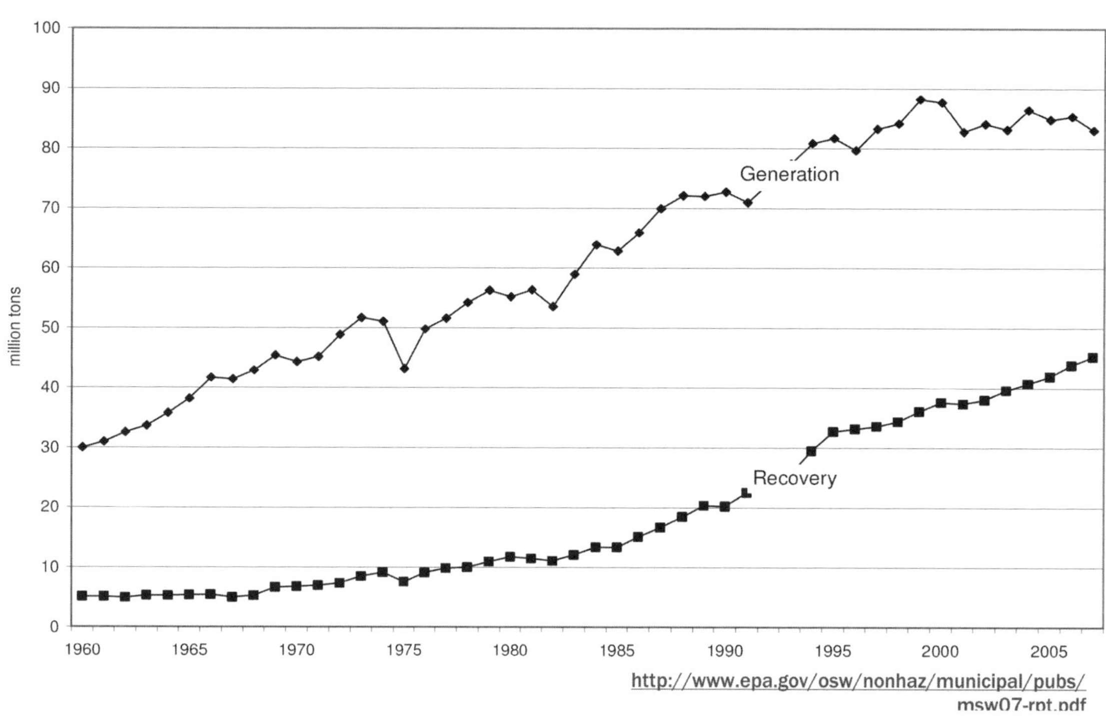

http://www.epa.gov/osw/nonhaz/municipal/pubs/msw07-rpt.pdf

Situation 4: Glass Wasting and Recycling

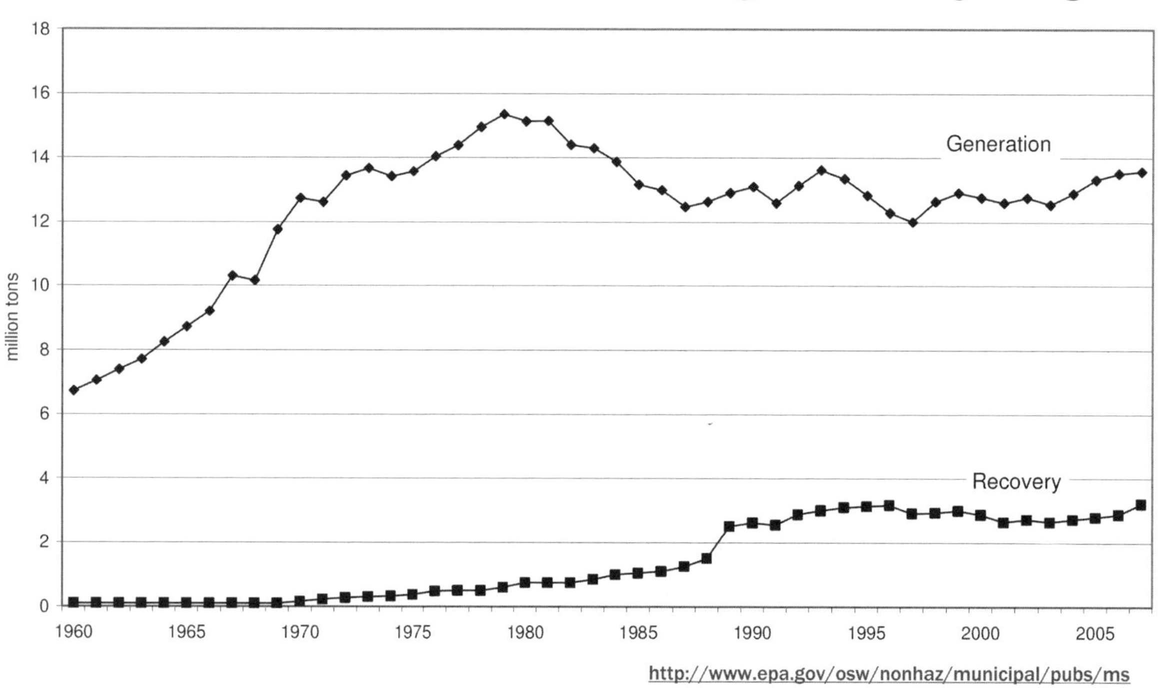

http://www.epa.gov/osw/nonhaz/municipal/pubs/msw07-rpt.pdf

Sentence Stems and Activities Aligned to Cross-Curricular Student Expectations

(subsection c)

Learning Strategies		
1(A) use prior knowledge and experiences to understand meanings in English	**Prior Knowledge** Teacher Questions • What do you know about...? • What do you remember about...? • What is an example of a...? • What is a ...? • What is a type of ...? • What experience have you had with...? • What have you learned about...? • Close your eyes and think of ___. What do you see? • What does the picture/word/phrase remind you of? • What does ___ mean to you? • How can ___ be represented? • How have you used...? • What do you think the word ___ means? • Think about a time when you... Student Sentence Stems • I know.... • I remember... • An example of a ___ is... • A ___ is... • A type of ___ is... • An experience I have had with ___ is... • I learned... • I see... • The picture reminds me of... • ___ means... • A ___ can be represented with a... • I have used ___ to... • I think ___ means... • A time I used ___ was when... Examples PK-5 • I know triangles have three sides. • I have used money to buy candy. • A learned perimeter is like a fence. 6-8 • An example of a graph is a circle graph. • I think translation means to change from one language to another. • An equation can be represented with a model. 9-12 • The pictures remind me of parallel lines. • I remember variables are used in equations. • A type of function is a linear function.	• Anticipation Chat • Anticipation Guides Insert Method • KWL • List/Group/Label • Pretest with a partner • Free Write • Math Sorts • Graffiti Write • Carousel Activity

Sentence Stems and Activities Aligned to Cross-Curricular Student Expectations

(subsection c)

<table>
<tr><th colspan="3">Learning Strategies</th></tr>
<tr>
<td rowspan="2">1(B) monitor oral and written language production and employ self-corrective techniques or other resources</td>
<td>Self Corrective Techniques
Teacher Question/Statements<ul><li>Please say that again.</li><li>Please repeat.</li><li>What I hear you say is...</li><li>Does the statement sound correct to you?</li><li>Does it sound right/look right?</li><li>What did you notice?</li><li>What letter does the word ___ begin with?</li><li>What sound does ___ make?</li></ul>Student Sentence Stems<ul><li>I mean ...</li><li>I meant to say/write ...</li><li>I said...</li><li>Let me rephrase that ...</li><li>Let me say that again ...</li><li>The word I am thinking of looks like...</li><li>How do you pronounce this word?</li><li>How would I be able to check ...?</li></ul></td>
<td rowspan="2"><ul><li>Accountable Conversation Questions</li><li>Oral Scaffolding</li><li>Think Alouds</li><li>Total Response Signals</li></ul></td>
</tr>
<tr>
<td>Examples
PK-5<ul><li>Student says, “I count three bear.” Teacher restates, “You counted three bears.” Student says, “Yes, I counted three bears.”</li><li>Student says, “The reminder is 5.” Then self-corrects by saying, “I mean the remainder is 5.”</li></ul>6-8<ul><li>I meant to write, “The radii of the circles are”... instead of “the radiuses of the circles are...”</li></ul>9-12<ul><li>How do you pronounce the word linear?</li></ul></td>
</tr>
<tr>
<td>1(C) use strategic learning techniques such as concept mapping, drawing, memorizing, comparing, contrasting, and reviewing to acquire basic and grade-level vocabulary</td>
<td>Drawing/Memorizing/Reviewing
Teacher Questions/Tasks<ul><li>What picture represents the word...?</li><li>What gesture could be used to represent...?</li><li>Create a rap song to remember...</li><li>Create a mnemonic to memorize...</li><li>Match the words to their corresponding picture.</li><li>What strategy could you use to remember the meaning of...?</li></ul></td>
<td><ul><li>CALLA Approach</li><li>Concept Definition Map</li><li>Concept Mapping</li><li>Creating Analogies</li><li>Creating Words</li><li>Flash Card Review</li><li>Four Corners Vocabulary</li><li>Graffiti Write</li><li>Math Sorts</li></ul></td>
</tr>
</table>

Sentence Stems and Activities Aligned to Cross-Curricular Student Expectations

(subsection c)

<table>
<tr><th colspan="3">Learning Strategies</th></tr>
<tr>
<td>1(C) continued</td>
<td>
Student Sentence Stems
<ul>
<li>I drew a ...</li>
<li>I can draw a ___ to represent a...</li>
<li>I can describe ___ by drawing a...</li>
<li>The model shows...</li>
<li>The picture represents the word...</li>
<li>The graph describes...</li>
<li>The symbol for ___ is...</li>
<li>The pattern is an example of a...</li>
<li>The word is ___ and it looks like this...</li>
<li>I memorized the ___ by remembering...</li>
<li>I decided to represent ___ this way because ...</li>
<li>I know/don't know the words ...</li>
<li>I'm familiar/not familiar with ___</li>
<li>I will need to review ...</li>
</ul>
Concept Mapping
Teacher Questions
<ul>
<li>What is the focus question of your concept map?</li>
<li>What important terms must you know to create a concept map of...?</li>
<li>What are the most important ideas in your concept map?</li>
<li>How does a concept map help you learn the meaning of...?</li>
<li>Why did you organize the information like that?</li>
</ul>
Student Sentence Stems
<ul>
<li>The focus question is ...</li>
<li>The terms I must know are...</li>
<li>The most important idea is...</li>
<li>___ is related to ___.</li>
<li>Some examples of a ___ are ...</li>
<li>A non-example is...</li>
</ul>
Comparing/Contrasting
Teacher Questions
<ul>
<li>What is the difference between...?</li>
<li>How are ___ and ___ the same?</li>
<li>Why do you think they are the same/different?</li>
</ul>
Student Sentence Stems
<ul>
<li>A ___ has ___.</li>
<li>A ___ is ___.</li>
<li>A similarity is...</li>
<li>A difference is...</li>
<li>A ___ has ___, but a ___ has ___.</li>
<li>___ and ___ both have ___.</li>
<li>An attribute ___ and ___ have in common is...</li>
<li>A property ___ and ___ don't share in common is...</li>
<li>___ is the same as ___ because they are both ...</li>
<li>___ is different from ___ because ...</li>
<li>___ is similar to ___ because ...</li>
<li>I think they are the same/different because...</li>
</ul>
</td>
<td>
<ul>
<li>Manipulatives</li>
<li>Multiple Representations</li>
<li>Graphic Organizer</li>
<li>Personal Dictionary</li>
<li>Scanning</li>
<li>Six Step Vocabulary Process</li>
<li>Total Physical Response (TPR)</li>
<li>Total Response Signals</li>
<li>Vocabulary Alive</li>
<li>Vocabulary Game Shows</li>
<li>Word Play</li>
<li>Word Sorts</li>
</ul>
</td>
</tr>
</table>

Sentence Stems and Activities Aligned to Cross-Curricular Student Expectations

(subsection c)

<table>
<tr><th colspan="3">Learning Strategies</th></tr>
<tr><td>1(C) continued</td><td>Examples
PK-5
• I can draw a <u>box</u> to represent <u>a cube</u>.
• The model shows a <u>fraction</u>.
• The most important idea in the concept map is <u>types of coins</u>.
• A <u>ruler</u> has <u>inches</u>. A <u>thermometer</u> has <u>degrees</u>.
6-8
• I memorized <u>order of operations</u> by remembering <u>Please Excuse My Dear Aunt Sally</u>.
• The word is <u>acute triangle</u> and it looks like this (student uses arms to create an acute triangle).
• <u>Integers</u> and <u>positive rational numbers</u> both have positive whole numbers.
9-12
• I can describe a <u>functional relationship</u> by drawing a man and woman getting married.
• <u>Direct variation</u> is related to <u>linear functions</u>.
• An attribute <u>linear functions</u> and <u>quadratic functions</u> have in common is <u>rate of change</u>.</td><td></td></tr>
<tr><td>1(D) speak using learning strategies such as requesting assistance, employing non-verbal cues, and using synonyms and circumlocution (conveying ideas by defining or describing when exact English words are not known)</td><td>Requesting Assistance
Teacher Questions
• Did you understand the question?
• Do you want me to repeat the question?
• What gesture can you use to tell me to speak slower?

Student Sentence Stems
• Can you help me ...?
• I don't understand ...
• Would you please repeat/rephrase that...?
• Would you please say that again a little slower?
• Would you please explain ...?

Synonyms/Circumlocution
Teacher Questions
• What other word can be used for the word ___?
• What is an example of a ___?
• What is a word that means the same as ___?
• Describe the word you are thinking about.
• What does the word/concept remind you of?

Student Sentence Stems
• It's the same as ...
• ___ means the same as ___.
• Another word for ___ is ___
• ___ is the same as the symbol ___.
• A synonym for the word ___ is ___.
• A ___ is a ___.
• It's similar to ...
• It includes ...
• The word I am thinking of looks like...</td><td>• CALLA Approach
• Accountable Conversation Questions
• Expert/Novice
• Instructional Scaffolding
• Think, Pair, Share
• Total Physical Response (TPR)
• Vocabulary Alive</td></tr>
</table>

Sentence Stems and Activities Aligned to Cross-Curricular Student Expectations

(subsection c)

<table>
<tr><th colspan="3">Learning Strategies</th></tr>
<tr><td>1(D) continued</td><td>Examples
PK-5
• Another word for <u>diamond</u> is <u>rhombus</u>.
• <u>Equal to</u> is the same as the symbol <u>=</u>.
6-8
• A synonym for the word <u>convert</u> is <u>change</u>.
• The graph I am thinking about looks like <u>a pizza</u>.
9-12
• Another word for <u>cross</u> is <u>intersect</u>.
• The word I am thinking of looks like <u>the letter u.</u></td><td></td></tr>
<tr><td>1(E) internalize new basic and academic language by using and reusing it in meaningful ways in speaking and writing activities that build concept and language attainment</td><td>Concept Attainment with New Words
Teacher Questions/Tasks
• How would you categorize the words/ pictures /numbers?
• What are the attributes of...?
• Classify the ....

Student Sentence Stems
• A characteristic is...
• A characteristic of ___ and ___ is...
• One characteristic/attribute of ___ is ...
• The group has...
• I would classify this word/concept under ___ category.
• The patterns have in common...
• All ___ are ...
• All ___ have ...
• All ___ are not ...
• All ___ do not have ...
• ___ is an example of ___ because...
• ___ is a non-example of ___ because ...
• Another example might be ___ because ....

Language Attainment with New Words
Teacher Questions/Tasks
• Share with a partner what you know about the word...
• Write words/examples associated with the word...
• Is there another word for ___?
• What math term have you learned that describes ___?
• What does ___ represent?
• Use the words ___, ___, and ___ in a complete sentence.

Student Sentence Stems
• ___ means ...
• ___ represents ___.
• A ___ is an attribute of a ___.
• Another word for ___ is ___.
• The ___ describes ...
• A math term that describes ___ is ___.
• I can use the word ___ when ...
• I would not use the word ___ when ...
• I might be able to use the word ___ when ___ because ...
• I probably would not use the word ___ when ___ because ...</td><td>• CALLA Approach
• Conga Line
• Concept Attainment
• Creating Words
• Concept Definition Map
• Conga Line
• Creating Analogies
• Dirty Laundry
• Fold the Line
• Group Response with a White Board
• Instructional Conversation
• Manipulatives
• Multiple Representations Graphic Organizer
• Order It Up Math Puzzle
• Question, Signal, Stem, Share, Assess
• Think, Pair, Share
• Whip Around
• Word Sorts
• Math Sorts</td></tr>
</table>

Sentence Stems and Activities Aligned to Cross-Curricular Student Expectations

(subsection c)

<table>
<tr><th colspan="3">Learning Strategies</th></tr>
<tr><td>1(E) continued</td><td>Examples
PK-5<ul><li>Another word for <u>bigger</u> is <u>greater than</u>.</li><li><u>Slide</u> describes <u>a translation</u>.</li></ul>6-8<ul><li>A math term that describes <u>a shape getting smaller</u> is <u>a dilation.</u></li><li><u>A 145° angle</u> is a non-example of an <u>acute angle</u>.</li></ul>9-12<ul><li>The <u>vertex of a parabola</u> describes the <u>high or low point of the graph</u>.</li><li>One characteristic of <u>an exponential function</u> is <u>it has no x-intercept</u>.</li></ul></td><td></td></tr>
<tr><td>1(F) use accessible language and learn new and essential language in the process</td><td>Using Accessible Language
Teacher Questions<ul><li>The word you are thinking of is...</li><li>What you are trying to say is...?</li><li>What I hear you say is...</li><li>Will you repeat this word with me?</li></ul>Student Sentence Stems<ul><li>I think...</li><li>The answer is...</li><li>The pattern is...</li><li>The operation is...</li><li>I need to say ...</li><li>To find out how to say __ I can look at ...</li><li>Will you please explain what ___ means?</li><li>I can use resources such as ___ to remember how to say ...</li></ul>Examples
PK-5<ul><li>I think <u>a peseta is more than a cent</u>. Teacher says, "What I hear you say is that a <u>quarter is more than a penny</u>."</li></ul>6-8<ul><li>If I want <u>to say the number that happens the most</u> I need to say <u>the mode</u>.</li></ul>9-12<ul><li>To find out how to say <u>a line is going up</u>, I can look at my notes.</li></ul></td><td><ul><li>Accountable Conversation Questions</li><li>CALLA Approach</li><li>Chat Room</li><li>Creating Words</li><li>Dirty Laundry</li><li>Expert/Novice</li><li>Instructional Scaffolding</li><li>Think Alouds</li><li>Vocabulary Alive</li></ul></td></tr>
</table>

Sentence Stems and Activities Aligned to Cross-Curricular Student Expectations

(subsection c)

<table>
<tr><th colspan="3">Learning Strategies</th></tr>
<tr><td>1(G) demonstrate an increasing ability to distinguish between formal and informal English and an increasing knowledge of when to use each one commensurate with grade-level learning expectations</td>
<td>Formal and Informal English
Teacher Questions
• What word could you use instead of ___?
• What word would you use to describe ___ to a friend?
• What word would you use to describe ___ to a math teacher?

Student Sentence Stems
• The ___ has ......
• I would describe ___ to a friend by saying...
• At school, we say ...
• I would describe this to someone in my own words by saying/writing ...
• I would describe that using mathematical language by saying ...

Examples
PK-5
• The triangle has 3 corners. VS. The triangle has 3 vertices.
• I did the problem. VS. I solved the problem.
6-8
• The shape flipped over this line. VS. The rectangle reflected through the x-axis.
• This is a number with repeating decimals. VS. This is a rational number.
• Numbers are going up. VS. Numbers are increasing.
9-12
• You minus the number on both sides of the problem. VS. Subtract the number on both sides of the equation.
• The y-values keep on going forever. VS. The range is all infinite numbers.</td>
<td>• Brick and Mortar Cards
• Discussion Starter Cards
• Formal/Informal Pairs
• Math Sorts
• Mix and Match
• Radio Talk Show
• Same Scene Twice
• Sentence Sort</td></tr>
<tr><td>1(H) develop and expand repertoire of learning strategies such as reasoning inductively or deductively, looking for patterns in language, and analyzing sayings and expressions commensurate with grade-level learning expectations</td>
<td>Deductive Reasoning
Teacher Questions
• What is given in the problem?
• The theorem says...
• If ___ is true, then we can state ...?

Student Sentence Stems
• All ___ have ___. ___ has ___. ___ must be a ___.
• All ___ measure ___. ___ measures ___. ___ must be ___.
• All ___ equations have a ___ and ___. ___ has ___ and ___. ___ must be an example of ___.
• All ___ have...
• So it must be an example of ...</td>
<td>• Instructional Conversation
• Perspective-Based Writing
• Question, Signal, Stem, and Share
• Structured Conversation</td></tr>
</table>

Sentence Stems and Activities Aligned to Cross-Curricular Student Expectations

(subsection c)

Learning Strategies		
1(H) continued	**Inductive Reasoning** Teacher Questions • What can you conclude from the pattern? • What observations did you make? • Why did you group these ___ together? • What is a characteristic all ___ have/had in common? Student Sentence Stems • My conjecture is... • A counterexample is... • A generalization is... • A conclusion is... • A conclusion I can make from the ___ is... • The function rule is... • All ___ had the characteristic(s) of... • If the pattern continues, I think it will... • All the ___ we saw were/had ... • So all ___ probably are/have... • Every example we observed was/had ... • So we can infer that all ___ are/have... • If ___ works for ___, maybe ___ will work for... **Patterns in Language** **Analyzing Sayings/Expressions** Teacher Question • What are words/expressions I often use? • What is the difference between an expression and a mathematical expression? • What is the difference between a sentence and a number sentence? Student Sentence Stems • The answer is... • The expression that best describes ___ is... • The equation that best represents ___ is... • The relationship between ___ and ___ is... • I could use the ___ number sentence to find the value of h. • ___ makes the equation true. • One pattern I noticed was ... • I think the word/expression ___ means... • One word/expression that was used a lot was...	
	Examples	
	PK-5 • Congruent figures have the same size same shape. These two figures have the same size same shape. The figures must be congruent. 6-8 • My conjecture is all those numbers are prime. 9-12 • One pattern I noticed was it starts at 3 and goes up by 5 each time.	

Sentence Stems and Activities Aligned to Cross-Curricular Student Expectations

(subsection c)

Listening		
2(A) distinguish sounds and intonation patterns of English with increasing ease	**Sounds and Intonation Patterns** Teacher Questions • Did I ask a question or make a statement? Why? • Can you tell ___ and ___ apart? How? • Can you distinguish between ___ and ___? • Why did I stress the word___? • Is the word ___ spelled with a ___ or a ___? Student Sentence Stems • What you said sounded like a ___ because... • Are you telling me or asking me? • I can tell ___ apart from ___. • I can distinguish between ___ and ___. • I can't distinguish between ___ and ___. • ___ sounds different than ___. • The word ___ and ___ sound the same to me. • You said the word ___. It starts with... • I think that word starts with the letter (is spelled) ___ because ... • You stressed the word ___ because ... • You did not stress the word ___ because ... • To change the meaning of this sentence I could stress ... • To change the tone of this sentence, I could (change the pitch, volume, speed, etc.) ... Examples PK-5 • You said the word kite. It starts with the letter K. 6-8 • The word mode and most sound the same to me. 9-12 • Teacher tells a student to verify if the answer is correct by stating, "That is the slope?" Student replies, "Are you telling me or asking me?"	• CCAP • Sound Scripting • Segmental Practice • Suprasegmental Practice
2(B) recognize elements of the English sound system in newly acquired vocabulary such as long and short vowels, silent letters, and consonant clusters	**Sound System** Teacher Questions • What sound does the letter(s) ___ make in the word ___? • What sound does the vowel ___ make in the word ___? • Is ___ a long/short vowel in the word ___? Student Sentence Stems • I hear ___ in the word ___. • The sound I heard was... • The word you said has a ... • ___ has the ___ sound. Examples PK-12 • The sound I heard was a short vowel "A". • The word range has a silent e. • Graph has the "f" sound.	• Word Wall • Word Sorts • Songs/Poems/ Rhymes • Systematic Phonics Instruction • Segmental Practice

Sentence Stems and Activities Aligned to Cross-Curricular Student Expectations

(subsection c)

<table>
<tr><th colspan="3">Listening</th></tr>
<tr><td rowspan="2">2(C) learn new language structures, expressions, and basic and academic vocabulary heard during classroom instruction and interactions</td><td>Language Structures/Expressions during Interactions
Teacher Question
• What new terms did you hear during the lesson?
• The following are some examples of question stems commonly used in TAKS. Students should get used to hearing these types of questions during instruction.
o Which expression best represents ___?
o Which expression could be used to___?
o Which equation best represents the ___?
o What is a common characteristic between ___ and ___?
o Which figure has exactly ___?
o If ___, what is ___?
o Which model best represents the ___?
o Which table best fits the ___?
o Which of these is the closest to ___?
o Which ___ appears to be ___?

Student Sentence Stems
• I heard the new word/phrase...
• One new phrase I used was ...
• I heard ___ use the word/phrase ...
• An expression I hear in math class is...
• A new word/phrase I heard was ...
• I can use that word/phrase when ...
• I used the word/phrase ___ when I spoke with ...
• I used the word/phrase ___ to express the idea that ...</td><td rowspan="2">• Carousel Activity
• Creating Words
• Oral Scaffolding
• Personal Dictionary
• Scanning
• Self-assessment of Levels of Word Knowledge
• Think, Pair, Share,
• Vocabulary Self Collection
• Vocabulary Alive
• Word Sorts</td></tr>
<tr><td>Examples
PK-5
• A new word I heard was <u>average</u>.
6-8
• An expression I hear in math class is "If <u>the pattern continues...</u>"
9-12
• A common sentence frame in geometry is, "If ___, then ___."</td></tr>
<tr><td>2(D) monitor understanding of spoken language during classroom instruction and interactions and seek clarification as needed</td><td>Clarification during Instruction and Interaction
Teacher Questions/Actions
• Do you need more time to think?
• Let me repeat the question.
• After asking a question, provide ELs sufficient wait-time to formulate a response before asking them to respond.

Student Sentence Stems
• Can you help me to ...?
• I don't understand what/how...
• Would you please repeat that?
• So you're saying ...
• May I please have some more information?
• May I have some time to think?</td><td>• Inside/Outside Circle
• Instructional Conversation
• Instructional Scaffolding
• Structured Conversation
• Think Alouds
• Think, Pair, Share
• Total Physical Response (TPR)</td></tr>
</table>

Sentence Stems and Activities Aligned to Cross-Curricular Student Expectations

(subsection c)

Listening		
2(E) use visual, contextual, and linguistic support to enhance and confirm understanding of increasingly complex and elaborated spoken language	**Linguistic, Visual, Contextual Support** Teacher Questions • Based on the pictures, what is the lesson going to be about? • How do manipulatives (color tiles/fraction bars/algebra tiles, etc.) help you understand math concepts? • How does graphing a table's data help you better understand? • How could you represent this pattern with manipulatives? • What is the problem situation about? • Is the problem situation continuous or discrete? • What graphic organizer would help you...? Student Sentence Stems • Color tiles help me understand... • Algebra tiles show... • The graphing calculator helps me... • The graphic organizer shows... • The word ___ is on the word wall. • Finding the area of the room helps me understand... • If I want to find out ___ I can ... • I can use ___ to check if I .... • When I hear ___ it tells me ... • Would you please show me on the... (diagram/picture/organizer/notes/etc.)....? Examples PK-5 • Color tiles help me understand perimeter. 6-8 • The graphic organizer shows how to represent a proportion in a table, graph, picture and equation. 9-12 • Will you please show me on the graphing calculator?	• Creating Words • Graphic Organizers • Inside/Outside Circle • Instructional Conversation • Instructional Scaffolding • Manipulatives • Multiple Representations Graphic Organizer • Multiple Representations Card Game • Nonlinguistic Representations • Posted Phrases and Stems • Structured Conversation • Think, Pair, Share • Graphing Calculator • Personal Dictionaries
2(F) listen to and derive meaning from a variety of media such as audio tape, video, DVD, and CD ROM to build and reinforce concept and language attainment	**Concept Attainment from a Variety of Media** Teacher Questions • How did the ___ help you understand ___? • How is ___ related to the concept of ___? • What was the___ about? • How does the math software help you understand ___? • How does the graphing calculator help you understand ___? Student Sentence Stems • I notice ... • ___ represents... • I conclude... • It appears... • I heard/saw a ... • ___ can be used in real life to... • The ___ was about... • The video on ___ helps me understand... • The song was about... • I heard/observed ___ which makes me think ... • I think ___ is an example of ___ because... • One characteristic/attribute of ____ that I heard/observed is ...	• Concept Attainment • Concept Mapping • Learning Logs and Journals • Chunking Input • Visual Literacy Frames • Pairs View

Sentence Stems and Activities Aligned to Cross-Curricular Student Expectations

(subsection c)

Listening		
2(F) continued	**Language Attainment from a Variety of Media** Teacher Questions • What words might you hear in the ___? • What new word did you hear/see in the ___? • What do you think ___ means? • Based on the ___, what do you think ___ means? • How does the graphing calculator help you understand the meaning of...? Student Sentence Stems • ___ means... • I heard/saw the word/phrase ___. • I think the word/ phrase means/does not mean ... • I heard/saw the word/phrase ___. I can use it when... • I heard/saw the word/phrase ___. I might be able to use it when___ because ... • I heard/saw the word/phrase___. I probably would not use it when ___ because ... Examples PK-5 • The PPT was about fact families. • I heard the words denominator and numerator. 6-8 • It appears the rap song is about adding fractions. • I heard the word proportion 3 times. 9-12 • The video on systems of equations helps me understand what the point of intersection means. • Y=1 in the graphing calculator means the first function.	

2(G) understand the general meaning, main points, and important details of spoken language ranging from situations in which topics, language, and contexts are familiar to unfamiliar	**Meaning in Spoken Language**
	Teacher Questions/Tasks
	• What was the ___ about?
	• What are the instructions?
	• Based on the explanation, how would you...?
	• Based on the clues given, what is the meaning of...?
	• Will you repeat what I just said?
	• What did you understand of what I said?
	• Prompt for elaborated student responses such as:
	  o Explain what ___ just said.
	  o Tell me more about that.
	  o Do you agree with ...? Why/why not?
	  o Why do you think...?
		• Dirty Laundry
		• Graffiti Write
		• Guess Your Corner
		• IEPT
		• Question, Signal, Stem, Share, Assess
		• Reciprocal Teaching
		• Story Telling
		• Structured Conversation
		• Summarization Frames

Sentence Stems and Activities Aligned to Cross-Curricular Student Expectations

(subsection c)

Listening		
2(G) continued	Student Sentence Stems • I think ___ means ... • I think ___means ___ because ... • The ___ is/is not represented by...because • My partner said... • I agree/disagree with ___ because... • ___ said "___." I think it means ... • I heard (the speaker) say ... • I heard you say___. Another way to say that might be... • One thing I heard was ... • One thing (the speaker) said was ... • I have never heard the words/phrase/concept but I think it means... **Main Point in Spoken Language** Teacher Questions • What steps must you follow to solve for...? • What is the word problem about? • What is the problem situation about? • What is the main point of the ___ property? • What information is irrelevant? Student Sentence Stems • The ___ is... • The ___ describes... • The ___ represents... • It's about... • The ___ property is about... • A generalization is... • I do/don't need to know about... • The word problem is about... • Overall our objective is... **Details in Spoken Language** Teacher Questions • What are the attributes/characteristics of...? • From a scale of 1 to 3, how clear were the instructions? • What is step number ___ of the instructions? • What is step number ___ to solve the ___? • What is one important detail you heard? • Why is ___ different from ___? • How are ___ different from ___? Student Sentence Stems • The instructions are... • First step is... • An attribute I heard is... • My partner said... • My partner said ___ and ___ have... • One characteristic of ___ is... • One important thing I heard (the speaker) say was ... • (The speaker) said ____, which is important because ... • I heard (the speaker) say ___ which supports the idea that ... • I heard that ___ and ___ are different because...	

Sentence Stems and Activities Aligned to Cross-Curricular Student Expectations

(subsection c)

Listening		
2(G) continued	**Examples** PK-5 • The teacher holds two yellow hexagons (pattern blocks) and says, "These two shapes are congruent." The student says, "I think congruent means the same." • The graph represents polygons and number of sides. • One characteristic of a picture graph is it has a title. 6-8 • I agree with my partner because integers don't have decimals. • The ratio of tires to cars is 4 to 1. • First step is to divide the paper in quarters. 9-12 • One thing I heard was b is the same as y-intercept. • Overall our objective is to graph square root functions. • The instructions are to represent $2x+3y=6$ in a graph, problem situation, table and picture.	
2(H) understand implicit ideas and information in increasingly complex spoken language commensurate with grade-level learning expectations	**Implicit Ideas** Teacher Questions • From the ___ what can you conclude? • What is your prediction? • Which procedure could be used for...? • Why statement is best supported by...? • Which of the following shows...? • Which statement is not true...? • Which is a reasonable answer? • What can you infer from the ___? • What conclusion can be drawn from...? • Based on the information in the ___, which statement is a valid conclusion? • Why might NOT be a valid representation of___? • Which is the best explanation for...? Student Sentence Stems • A valid conclusion is... • I think ___ because ... • The statement is not true because... • I can conclude ___ because... • I can assume ____ because ... • The best explanation is... • My prediction is... • Even though it doesn't say ___, I think ... • Based on ___, I can infer that ... • From the information found in ___ I can infer that ___ because ... • From the ___, I can infer... • Based on the information I heard in ___, I can conclude... **Examples** PK-5 • I think I need more base-ten blocks because you said the next number is going to be larger. 6-8 • The statement is not true because you said irrational numbers have non-repeating and non-terminating decimals. 9-12 • I can conclude the parabola will face down because you said the equation is $y=-3x^2$.	• Instructional Conversation • Discovery Learning • Order It Up Math Puzzle • Question, Signal, Stem, Share, Assess • Reciprocal Teaching • Story Telling • Structured Conversation • Summarization Frames • Whip Around

Sentence Stems and Activities Aligned to Cross-Curricular Student Expectations

(subsection c)

Listening		

2(I) demonstrate listening comprehension of increasingly complex spoken English by following directions, retelling or summarizing spoken messages, responding to questions and requests, collaborating with peers, and taking notes commensurate with content and grade-level needs	**Following Spoken Directions**
	Teacher Questions/Tasks
	• Who can restate the instructions?
	• What will we do first, second, and finally?
	• Tell your partners what they need to do to complete the task.
	
	Student Sentence Stems
	• The first step is ...
	• The next steps are ...
	• I know I'm finished when ...
	• What you need to do is...
	• The initial step is ...
	• The next step(s) in the process is/are
	• I know I've completed the task successfully when ...
	
	Retelling/Summarizing Spoken English
	Teacher Questions/Tasks
	• In your own words, what did ___ just say?
	• Prompt for elaborated student responses such as:
	    ○ Explain what ___ just said.
	    ○ Tell me more about that.
	    ○ Do you agree with ...? Why/why not?
	    ○ Why do you think...?
	
	Student Sentence Stems
	• It's about...
	• The main idea is ...
	• First.... Then.... Finally...
	• I would explain the concept to a friend by ...
	• The general idea is...
	
	Responding to Questions/Requests
	Teacher Questions
	• Did anyone think of this problem in a different way?
	• Why did you choose this method?
	• How does the method relate to the method ___ just explained?
	• Why did you draw ___ to represent a ___?
	• ___ please come and solve problem number ___ on the board.
	
	Student Sentence Stems
	• You asked___. I think ...
	• The answer is...
	• I think you're asking ...
	• Do you want me to ...?
	• I heard you say___, so I need to ....
		• Carousel Activity
		• Creating Words
		• Dirty Laundry
		• Guess Your Corner
		• Framed Oral Recap
		• Keep, Delete, Substitute, Select
		• IEPT (Peer Tutoring)
		• Instructional Conversation
		• Mix and Match
		• Note Taking Strategies
		• Outlines
		• Question Answer Relationship (QAR)
		• Question, Signal, Stem, Share, Assess
		• Reader/Writer/Speaker Response Triads
		• Reciprocal Teaching
		• Story Telling
		• Structured Conversation
		• Summarization Frames
		• Tiered Questions
		• Tiered Response Stems
		• W.I.T. Questioning
		• Word MES Questioning

Sentence Stems and Activities Aligned to Cross-Curricular Student Expectations

(subsection c)

<table>
<tr><th colspan="3">Listening</th></tr>
<tr>
<td>2(I) continued</td>
<td>
Collaborating With Peers
Teacher Questions
<ul>
<li>Grouping configurations involving ELLs needs to be predetermined prior to the beginning of instruction. Consider the following when grouping ELLs:
<ul>
<li>What is the purpose for grouping students?</li>
<li>What are the language proficiency levels and language backgrounds of students?</li>
<li>Does the grouping configuration(s) meet the lesson's objectives?</li>
</ul></li>
</ul>
Student Sentence Stems
<ul>
<li>Can you help me understand ...?</li>
<li>Would you please repeat that?</li>
<li>What do you think...?</li>
<li>Who's responsible for...?</li>
<li>Who should ...?</li>
<li>My job/part/role is to...</li>
<li>So I should ...</li>
<li>I'm responsible for ...</li>
<li>First ___. Second ___. Finally ___.</li>
</ul>
Taking Notes
Teacher Questions
<ul>
<li>What information did you write down?</li>
<li>How did you organize the information? Why?</li>
</ul>
Student Sentence Stems
<ul>
<li>I noted ...</li>
<li>The main ideas I wrote down were ...</li>
<li>Some details I wrote down were ...</li>
<li>I can organize the ideas I wrote by... (making an outline, concept map, Venn diagram, chart, etc.)</li>
</ul>
</td>
<td></td>
</tr>
<tr><th colspan="3">Speaking</th></tr>
<tr>
<td>3(A) practice producing sounds of newly acquired vocabulary such as long and short vowels, silent letters, and consonant clusters to pronounce English words in a manner that is increasingly comprehensible</td>
<td>
Producing Sounds
Teacher Questions
<ul>
<li>What sound does/do the letter(s) ___ make in the word ___?</li>
<li>What sound does the vowel ___ make in the word ___?</li>
<li>Which word has the consonant blend ___?</li>
<li>Is ___ a long/short vowel in the word ___?</li>
<li>How would you pronounce the word ___?</li>
</ul>
Student Sentence Stems
<ul>
<li>___ makes the ___ sound.</li>
<li>___ is pronounced ___.</li>
<li>The word ___ is pronounced ___ because ...</li>
<li>The letter(s) ___ make(s) the ___ sound.</li>
<li>The word ___ begins with the letter...</li>
<li>The word ___ has the long/short vowel ...</li>
<li>The word ___ has a silent ...</li>
<li>The word ___ has the consonant blend ...</li>
<li>The letter ___ in the word ___is long because ...</li>
<li>The ___ is silent in the word ___ because...</li>
</ul>
</td>
<td>
<ul>
<li>Fluency Workshop</li>
<li>List Stressed Words</li>
<li>Recasting</li>
<li>Segmental Practice</li>
<li>Suprasegmental Practice</li>
</ul>
</td>
</tr>
</table>

Sentence Stems and Activities Aligned to Cross-Curricular Student Expectations

(subsection c)

Speaking

3(A) continuted

Examples

PK-5
- The word eight has the silent letters gh.

6-8
- The word (teacher holds up an index card with the word Pythagorean) is pronounced Pythagorean.

9-12
- The letters PH make an "f" sound in the word graphical.

3(B) expand and internalize initial English vocabulary by learning and using high-frequency English words necessary for identifying and describing people, places, and objects, by retelling simple stories and basic information represented or supported by pictures, and by learning and using routine language needed for classroom communication

Description and Simple Story Telling with High Frequency Words and Visuals

Examples of high frequency words

About	Better	Family	Little	Picture
Above	Between	Feet	Long	Place
All	Big	Few	Many	Point
Almost	Both	Find	Money	Right
Also	Change	First	Move	Same
Always	Different	If	Next	Sentence
Answer	Down	Important	Not	Second
Around	Enough	Large	Number	Small
Below	Example	Learn	Often	Time
Best	Face	Left	Only	Together

- My picture is about…
- ___ looks like…
- I can describe ___ with the words…
- The picture(s) show(s) …
- I know it is a ___ because…
- ___ could be described as___ because …
- I can draw a ___ to represent a ___.
- A model of a ___ will help me tell you…

Routine Language for Classroom Communication

Teacher Questions
- What gesture do I use to let you know to get into groups?
- If I raise my right hand it means…?
- If I say … it means it is time for…?
- If you don't understand what I am saying, you can say..?

Student Sentence Stems
- ___ means___.
- Where is/are…?
- Where do I…?
- How do I …?
- Can you help me ….?
- May I please have some more information?
- May I ask someone for help?
- May I go to…?
- May I…?
- When is it time to …?

Activities:
- Accountable Conversation Questions
- Conga Line
- Dirty Laundry
- Expert/Novice
- Inside/Outside Circle
- Instructional Conversation
- Numbered Heads Together
- Partner Reading
- Question, Signal, Stem, Share Assess
- Retelling
- Summarization Frames
- Think, Pair, Share
- Vocabulary Alive

Sentence Stems and Activities Aligned to Cross-Curricular Student Expectations

(subsection c)

<table>
<tr><th colspan="3">Speaking</th></tr>
<tr>
<td>3(C) speak using a variety of grammatical structures, sentence lengths, sentence types, and connecting words with increasing accuracy and ease as more English is acquired</td>
<td>
Speak using a variety of Structures

Teacher Questions
• Orally explain...
  ○ What are the attributes of...?
  ○ How would you order ___ from ___ to ___?
  ○ What will happen if...?
  ○ What are the similarities between ___ and ___?
  ○ Predict what the next ___ will be.
  ○ What can you infer from the ___?
  ○ What can you conclude from ___?

Student Sentence Stems
Description
• A ___ has ___.
• A ___ has ___, ___, and ___.
• A ___ is ___, ___, and ___.
• Additionally ___ has ...
• ___ is an example of...
• ___ is an example of.... because ...
Sequence
• First___. Second___.
• First ___ and then...
• If numbers ___ are ordered from ___ to ___, the first number would be...
Cause and Effect
• The ___ is ___.
• ___ because ___.
• The cause is ___. The effect is ___.
• ___ was caused by ___.
• If ___, then ___.
• When ___, then...
• The independent variable is ___, and the dependent variable is___.
Comparison
• A ___ has ___.
• A ___ has ___ but a ___ has ___.
• ___ and ___ both have...
• ___ is the same as ___.
• ___ differs from ___ in that...
• Although ___ has ___, ___ has ___.
• ___on the other hand has...
Predictions
• The ___ will have...
• The ___ will be...
• I predict ___ will...
• I predict ___ will ___ because...
• The next pattern will be...
• Due to ___, I think ___ will happen.
• Consequently, I think...
Inferences
• I can infer that...
• I know ___ because...
• My conjecture is...
• From the ___, I can infer...
</td>
<td>
• Canned Questions
• Conga Line
• Instructional Conversation
• Experiments/Lab
• Discovery Learning
• Fold the Line
• Numbered Heads Together
• IEPT (Peer Tutoring)
• Question, Signal, Stem, Share, Assess
• Reader/Writer/ Speaker Response Triads
• Signal Words
• Story Telling
• Structured Conversation
• Summarization Frames
</td>
</tr>
</table>

Sentence Stems and Activities Aligned to Cross-Curricular Student Expectations

(subsection c)

<table>
<tr><th colspan="3">Speaking</th></tr>
<tr><td rowspan="2">3(C) continued</td><td>Conclusion
• All ___ are ___.
• ___ are ___.
• I concluded...
• I can conclude that...
• If ___, then ___. Therefore...</td><td rowspan="3"></td></tr>
<tr><td>Examples
PK-5
Description
• A pentagon has sides.
• A pentagon has 5 sides, 5 vertices, and 5 angles.
Sequence
• First you add. Second you subtract.
• First you add and then subtract.
6-8
Cause and Effect
• The pattern is getting bigger.
• The pattern is getting bigger because you add three color tiles each time.
Comparison
• A bar graph has bars.
• A bar graph has bars but a circle graph has sectors.
Prediction
• The marble will be red.
• I predict the next marble will be red because there are more red than blue marbles.
9-12
Inferences
• I infer that the graph will not cross the x-axis.
• My conjecture is that the pattern represents the function y=2x+1..
Conclusion
• All functions are relations.
• I concluded that problem situations can be discrete or continuous.</td></tr>
<tr><td>3(D) speak using grade-level content area vocabulary in context to internalize new English words and build academic language proficiency</td><td>Speak using Math Vocabulary
Teacher Questions
The following are some examples of question stems commonly used in TAKS. It is recommended for students to use terms found in these questions during oral discussions.
• Which expression best represents ___?
• Which expression could be used to___?
• Which equation best represents the ___?
• What is the equation for ___?
• What is a common characteristic between ___ and ___?
• Which figure has exactly ___?
• If ___, what is ___?
• Which ___ best represents ___?
• Which table best fits the ___?
• Which of these is the closest to ___?
• Which ___ appears to be ___?
• Which ___ describes ___?
• What is the value of___?
• Which ___ can be used to determine ___?
• Which of the following statements is always true?
• What is the effect of ___ when ___ changes?</td><td>• Content Specific Stems
• Creating Analogies
• Creating Words
• Dirty Laundry
• Instructional Conversation
• Mix and Match
• Self-assessment of Levels of Word Knowledge
• Structured Conversation
• Question, Signal, Stem, Share, Assess
• Reciprocal Teaching</td></tr>
</table>

Sentence Stems and Activities Aligned to Cross-Curricular Student Expectations

(subsection c)

Speaking		
3(D) continued	Student Sentence Stems • ___ best represents.___. • ___ could be used to___. • ___ is for ___. • A common characteristic between ___ and ___ is... • ___ has exactly ___. • If ___, then ___. • ___ best fits ___. • ___ is the closest to ___. • ___ appears to be ___. • ___ describes ___. • The solution is... • The effect is... • The value of ___ is... • ___ can be used to determine ___. • The ___ that is always true is... • ___ will make the equation true. • If the pattern continues, it will take... • At this rate, it will take...	
	Examples of Math Vocabulary	

PK-5				
Add	Array	Acute angle	Clock	Bar graph
Columns	Additive pattern	Circle	Degree	Bar type graph
Cost	Equation	Closed figures	Distance	Chance
Decimal	Generate	Congruent	Dozen	Data
Divide	Number pairs	Edge	Elapsed time	Line graph
Dollar	Number Sentence	Line of symmetry	Gram	Mean
Equivalent	Paired numbers	Parallel lines	Half hour	Median
Fraction	Pattern	Polygon	Measure	Mode
Money	Relationship	Translation	Thermometer	Pictograph
Subtract	Table	Vertex	Yard	Possible outcomes
6-8				
Divide Fractions	Constant rate	Complimentary angles	Capacity	Circle graph
Exponent	Equation	Congruent	Centimeter	Complement
GCF	Expression	Coordinate plane	Circumference	Independent event
Integer	Proportion	Dilation	Customary	Possible combinations
Irrational number	Range	Obtuse angle	Mass	Probability
LCM	Ratio	Pythagorean theorem	Metric	Random
Multiply fractions	Table	Right angle	Mile	Sample space
Percent	Term number	Right triangle	Pi	Simple event
Prime factorization	Unit rate	Scalene triangle	Surface area	Stem and leaf plot
Rational numbers	Variables	Transformation	Volume	Venn diagram
9-12				
Absolute value function	Commutative property	Discrete	Laws of exponents	Sine
Alternate interior angles	Completing the square	Discrete situation	Linear function	Slope
Angle of depression	Composite figure	Distance formula	Logarithm	Square root function
Apothem	Conic sections	Distributive property	Midpoint formula	Systems of equations
Arc length	Conjecture	Domain	Monomial	Tangent
Associative property	Continuous situation	Ellipses	Parabola	Tangent line
Asymptote	Contrapositive	Exponential function	Polynomial	Trinomial
Axiomatic system	Converse	Geometric sequence	Quadratic function	X-intercept
Binomial	Cosine	Hyperbolas	Range	Y-intercept
Central angle	Dependent variable	Independent variable	Rate of change	Zero pairs

Sentence Stems and Activities Aligned to Cross-Curricular Student Expectations

(subsection c)

<table>
<tr><th colspan="3">Speaking</th></tr>
<tr>
<td>3(E) share information in cooperative learning interactions</td>
<td>Share in Cooperative Interactions
Teacher Questions
<ul>
<li>Grouping configurations involving ELLs needs to be predetermined prior to the beginning of instruction. Consider the following when grouping ELLs:
<ul>
<li>What is the purpose for grouping students?</li>
<li>What are the language proficiency levels and language backgrounds of students?</li>
<li>Does the grouping configuration(s) meet the lesson's objectives?</li>
</ul></li>
</ul>
Student Sentence Stems
<ul>
<li>An idea is...</li>
<li>My guess is...</li>
<li>I think...</li>
<li>A characteristic is...</li>
<li>First... Second... Finally...</li>
<li>The way I would solve the problem is...</li>
<li>The ___ can be represented with...</li>
<li>What I know about ___ is...</li>
<li>My suggestion would be ___ because...</li>
<li>I agree/disagree that...because...</li>
<li>In my opinion, the answer is reasonable because</li>
<li>To solve the problem, we can...</li>
<li>Is your answer reasonable? How do you know?</li>
</ul></td>
<td><ul>
<li>Carousel Activity</li>
<li>Conga Line</li>
<li>Fold the Line</li>
<li>Inside Outside Circle</li>
<li>Instructional Conversation</li>
<li>Structured Conversation</li>
<li>Question, Signal, Stem, Share, Assess</li>
<li>Peer Editing</li>
<li>Pairs View</li>
<li>Partner Reading</li>
<li>Interview Grids</li>
</ul></td>
</tr>
<tr>
<td>3(F) ask and give information ranging from using a very limited bank of high-frequency, high-need, concrete vocabulary, including key words and expressions needed for basic communication in academic and social contexts, to using abstract and content-based vocabulary during extended speaking assignments</td>
<td>Ask and Give Information
Teacher Questions/Requests
<ul>
<li>Write two questions for your partner to answer.</li>
<li>What is a question you have about the lesson?</li>
<li>What is your answer to the question...?</li>
<li>Explain your reasoning to a partner.</li>
<li>Explain to your partner why you agree/disagree with his answer?</li>
</ul>
NOTE: How, what, why, where, and when are high frequency words.

Student Sentence Stems
<table>
<tr><th>Ask for Information</th><th>Give Information</th></tr>
<tr><td>How do you ...?</td><td>First you ... then...</td></tr>
<tr><td>What is...?</td><td>___ is ...</td></tr>
<tr><td>What did you notice about/in...?</td><td>I noticed ...</td></tr>
<tr><td>What are the attributes of ...?</td><td>One attribute of ___ is...</td></tr>
<tr><td>What do you think caused ...?</td><td>I think ___ caused ___ because...</td></tr>
<tr><td>When do you...?</td><td>You ___ when....</td></tr>
<tr><td>Where do you place...?</td><td>You ___ the ___ in the...</td></tr>
<tr><td>Why did you solve...?</td><td>I solved it this way because...</td></tr>
</table></td>
<td><ul>
<li>Instructional Conversation</li>
<li>Interview Grids</li>
<li>Mix and Match</li>
<li>Order It Up Math Puzzle</li>
<li>Question, Signal, Stem, Share, Assess</li>
<li>Structured Conversation</li>
<li>Think, Pair, Share</li>
</ul></td>
</tr>
</table>

Sentence Stems and Activities Aligned to Cross Curricular Student Expectations

(subsection c)

Speaking		
3(G) express opinions, ideas, and feelings ranging from communicating single words and short phrases to participating in extended discussions on a variety of social and grade-appropriate academic topics	**Express Opinions, Ideas, and Feelings** Teacher Questions • What do you think about...? • What is your position on...? • Is the answer reasonable? Why? • How did you reach that solution/conclusion? • Will you please elaborate on your response? • Do you agree/disagree with...? Why? • Is there another solution to this problem? Please explain. • Is there a counterexample to ....? State the counterexample. • Tell me more about... • What else can you tell me about...? Student Sentence Stems • I believe ___. • My position is ___. • I think ___. • I think ___ because... • In addition, I think... • I predict... • A solution is.... • I solved the problem by... • The problem can be solved by... • Another solution is ___ because... • The answer is ___. • The answer is ___ because... • The answer is/isn't reasonable because... • I agree/disagree because... • I agree/disagree with ___ because... • ___ represents a ___. • ___ represents a ___ because... Examples PK-5 • I think the pattern will get bigger. • I think the pattern will grow by three tiles because it grew by three in the previous two patterns. 6-8 • The graph represents a translation. • The graph represents a horizontal translation because the figure moved three units to the right. 9-12 • The problem can be solved by using Cramer's rule. • I agree because fractions can't represent the domain.	• **Anticipation Chat** • **Conga Line** • **Instructional Conversation** • **Question, Signal, Stem, Share, Assess** • **Reciprocal Teaching** • **Structured Conversation** • **Think, Pair, Share** • **W.I.T. Questioning**

Sentence Stems and Activities Aligned to Cross-Curricular Student Expectations

(subsection c)

Speaking		
3(H) narrate, describe, and explain with increasing specificity and detail as more English is acquired	**Narrate, Describe, and Explain with Increasing Detail** Teacher Questions • How would you describe...? • In your own words, explain why... • Why did ___ happen? • What else can you say about...? • Will you please restate what ___ said? • Explain how you got that solution. Student Sentence Stems • ___ is ___. • This is a ... • This is a... and it has/is___ and ___. • The solution is... • The solution is... because... • ___ best represents ___. • ___ best represents ___ because... • ___ is about... • The most important attribute is... • ___ is the most important attribute because... • It's important to remember... • Initially....then... ultimately... • First..., then..., finally... Examples PK-5 • This is a clock. • This is a clock and it has minutes and hours. 6-8 • The solution is 5. • The solution is 5 because two squared plus three is seven and seven minus two is five. 9-12 • The most important attribute is rate of change. • The most important attribute is the slope because it is always increasing at a constant rate.	• Creating Words • Instructional Conversation • Numbered Heads Together • Question, Signal, Stem, Share, Assess • Roundtable • Story Telling • Structured Conversation • Summarization Frames
3(I) adapt spoken language appropriately for formal and informal purposes	**Formal and Informal Spoken English** • ___ means ___. • ___ could be a ___ but in math ___ means ___. • The object... • Another word for ___ is ___. • At school we say ___ instead of... • I would explain the pattern/table/graph/picture to a friend by ... • In math we use the word/phrase ... to ... • I would describe ___ to someone outside of school by ... • I would describe ___ using math language by ... Examples PK-5 • The object moved up. • The object translated vertically. 6-8 • Another word for big is enlarged. • At school, we say rectangular prism or cube instead of box. 9-12 • In algebra, we use the word function to describe how y is dependent on x.	• Chat Room • Expert/Novice • Mix and Match • Oral Scaffolding • Radio Talk Show • Sentence Sort • Word Sorts

Sentence Stems and Activities Aligned to Cross-Curricular Student Expectations

(subsection c)

Speaking		
3(J) respond orally to information presented in a wide variety of print, electronic, audio, and visual media to build and reinforce concept and language attainment	**Concept Attainment from a Variety of Media** Teacher Questions • How did the ___ help you understand ___? • How is ___ related to the concept of ___? • What was the___ about? • How does the math software help you understand ___? • How does the graphing calculator help you understand ___? Student Sentence Stems • I notice ... • ___ represents... • I concluded... • It appears... • I heard/saw a ... • ___ can be used in the real life to... • The ___ was about... • The video on ___ helps me understand... • The song was about... • I heard/observed ___ which makes me think ... • I think ___ is an example of ___ because... • One characteristic/attribute of ____ that I heard/observed is ... **Language Attainment from a Variety of Media** Teacher Questions • What words might you hear in the ___? • What new word did you hear/see in the ___? • What do you think ___ means? • Based on the ___, what do you think ___ means? • How does the graphing calculator help you understand the meaning of...? • What did they mean by...? • Why did they use the word ___ to describe ___? Student Sentence Stems • ___ means... • I think the word means/does not mean ... • I see/hear... • The word ___ was used... • I noticed the word ___ is pronounced... • I heard/saw the word(s) ___. • I heard/saw the word ___. I can use it when... • Words that were unfamiliar are ___, ___ and ___.	• Chunking Input • Concept Attainment • Concept Definition Map • Graphing Calculator • Learning Logs and Journals • Manipulatives • Pairs View • Visual Literacy Frames
	Examples PK-5 • The PPT was about parallel and perpendicular lines. • I heard the words repeating pattern. 6-8 • It appears that when you multiply fractions the product gets smaller. • I noticed the word radius is pronounced radius. 9-12 • The video on systems of equations helps me understand what the point of intersection means. • The button (-) in the graphing calculator means negative sign..	

Sentence Stems and Activities Aligned to Cross-Curricular Student Expectations

(subsection c)

Reading

4(A) learn relationships between sounds and letters of the English language and decode (sound out) words using a combination of skills such as recognizing sound-letter relationships and identifying cognates, affixes, roots, and base words

Decoding

Teacher Questions

- What sound does/do the letter(s) ___ make in the word ___?
- What sound does/do the vowel(s) ___ make in the word ___?
- Is ___ a long/short vowel in the word ___?

Student Sentence Stems

- The letter(s) ___ make(s) the ___ sound...
- The word ___ has the long/short vowel ...
- The word ___ has a silent ...
- The word ___ has the consonant blend ...
- The letter ___ in the word ___is long because ...
- The ___ is silent in the word ___ because...
- The word ___ is pronounced ___ because ...

Cognates

Teacher Questions

- What is a cognate?
- What is the cognate of the word...?
- What are some examples of false cognates?

Student Sentence Stems

- The word ___ helps me spell the word ___.
- The word ___ sounds like ___ in my language and means ...
- The word ___ sounds like___ in my language, but does NOT mean...

Examples of Math Cognates

Angle Ángulo	Cube Cubo	Digit Dígito	Factors Factores	Gallon Galón
Mode Modo	Number Número	Origin Origen	Pint Pinta	Probability Probabilidad
Radius Radio	Range Rango	Solution Solución	Sum Suma	Term Término

Affixes, Roots, and Base words

Teacher Questions

- What does the prefix/suffix ___ mean?
- What does the root word ___ mean?
- How does knowing the meaning of ___ help you figure out what ___ means?

Student Sentence Stems

- ___ means ___.
- ___ means ___ because...
- The word___ has the prefix/suffix/root ___ which means...
- The base word is...
- The base word in the word ___ is...
- Some other words with this prefix/suffix/root are ...
- This word probably means ___ because...

Examples of Roots, Base Words, Prefixes, and Suffixes used in Math

Root	Base Word	Prefix	Suffix
Circum	Angle	In-	-ed
Equi	Estimate	Tri-	-tion
Poly	Express	Under-	-sion

- Direct Teaching of Affixes
- Direct Teaching of Cognates
- Direct Teaching of Roots
- Self-assessment of Levels of Word Knowledge
- Word Generation
- Word Sorts
- Word Study Books
- Word Walls

Sentence Stems and Activities Aligned to Cross-Curricular Student Expectations

(subsection c)

<table>
<tr><th colspan="3">Reading</th></tr>
<tr>
<td>4(B) recognize directionality of English reading such as left to right and top to bottom</td>
<td>
Directionality of English Text

Teacher Questions
<ul>
<li>What is the directionality of script of ___ language?
<ul>
<li>Arabic (Right to Left)</li>
<li>English (Left to Right)</li>
<li>Korean (Left to Right or Top to Bottom)</li>
<li>Mandarin (Left to Right or Top to Bottom)</li>
<li>Russian (Left to Right)</li>
<li>Urdu (Right to Left)</li>
</ul></li>
</ul>
Student Sentence Stems
<ul>
<li>In English, words go ... (students can use gestures to indicate directionality)</li>
<li>In ____ (Chinese/Arabic/Hebrew etc.) words go..., but in English words go...</li>
<li>In___ (Spanish/French/Russian etc.) words go...., and in English words also go...</li>
</ul>
</td>
<td>
<ul>
<li>Total Physical Response (TPR)</li>
<li>Directionality Sort</li>
</ul>
</td>
</tr>
<tr>
<td>4(C) develop basic sight vocabulary, derive meaning of environmental print, and comprehend English vocabulary and language structures used routinely in written classroom materials</td>
<td>
Sight Vocabulary/

Teacher Questions:
<ul>
<li>How many times did you use/read/hear the sight word ___?</li>
</ul>
Examples of sight vocabulary words
<table>
<tr><td>A</td><td>Be</td><td>Fast</td><td>May</td><td>To</td></tr>
<tr><td>And</td><td>Call</td><td>Give</td><td>Some</td><td>Use</td></tr>
<tr><td>All</td><td>Cut</td><td>Going</td><td>The</td><td>We</td></tr>
<tr><td>Are</td><td>Every</td><td>Had</td><td>Think</td><td>You</td></tr>
</table>
Student Sentence Stems
<ul>
<li>I know...</li>
<li>___ means...</li>
<li>I used/read/heard the word ___.</li>
<li>What are sight words?</li>
</ul>
Environmental Print
Teacher Questions
<ul>
<li>What is environmental print?</li>
<li>How does environmental print help students comprehend vocabulary?</li>
<li>What should be labeled in the classroom?</li>
<li>Students show understanding of environmental print through actions with gestures or use simple phrases.</li>
</ul>
Student Sentence Stems
<ul>
<li>This is a ___.</li>
<li>The symbol says/means...</li>
<li>This sign says ____. It tells me...</li>
<li>Labeling things in the room helps me understand____because...</li>
</ul>
Examples
PK-12
Student points to posted illustration and says...
<ul>
<li>This is a <u>circle</u>.</li>
<li>The symbol means <u>equal to</u>.</li>
<li>That is an <u>ellipse</u>.</li>
</ul>
</td>
<td>
<ul>
<li>Expert/Novice</li>
<li>Oral Scaffolding</li>
<li>Total Physical Response (TPR)</li>
</ul>
</td>
</tr>
</table>

Sentence Stems and Activities Aligned to Cross-Curricular Student Expectations

(subsection c)

<table>
<tr><th colspan="3">Reading</th></tr>
<tr>
<td>4(D) use pre-reading supports such as graphic organizers, illustrations, and pretaught topic-related vocabulary and other prereading activities to enhance comprehension of written text</td>
<td>Pre Reading Supports
Teacher Questions
<ul>
<li>What do you know about...?</li>
<li>What do you remember about...?</li>
<li>What is an example of a...?</li>
<li>What is a ...?</li>
<li>What is a type of ...?</li>
<li>What experience have you had with...?</li>
<li>What have you learned about...?</li>
<li>Close your eyes and think of ___. What do you see?</li>
<li>What comes to mind when you think of...?</li>
<li>What does the picture/word/phrase remind you of?</li>
<li>What does ___ mean to you?</li>
<li>How can ___ be represented?</li>
<li>How have you used...?</li>
<li>What do you think the word ___ means?</li>
<li>What do you know about...? What do you want to learn about...?</li>
<li>Is ___ true or false?</li>
<li>What words in the story problem are unfamiliar to you?</li>
<li>Think about a time when you...</li>
</ul>
Student Sentence Stems
<ul>
<li>I know....</li>
<li>I remember...</li>
<li>A ___ is...</li>
<li>A type of ___ is...</li>
<li>I learned...</li>
<li>I see...</li>
<li>I have used ___ to...</li>
<li>An example of a ___ is...</li>
<li>The pictures are about ...</li>
<li>The statement is (true/false).</li>
<li>I think ___ means...</li>
<li>I think this ___ is about ...</li>
<li>A time I used ___ was when...</li>
<li>An experience I have had with ___ is...</li>
<li>A ___ can be represented with a...</li>
<li>The diagram/table/graph helps me...</li>
<li>The graphic organizer is about...</li>
<li>The organizer shows me that ___ is significant because ...</li>
<li>The diagram/table/graph tells me the ___ is about ...</li>
<li>The strategy that will help me understand this word problem the best is probably.... (note taking, scanning, surveying key text features such as bold words, illustrations and headings, using the wordlist, etc.)</li>
</ul></td>
<td><ul>
<li>Advance Organizers</li>
<li>Anticipation Guides</li>
<li>Backwards Book Walk</li>
<li>Brainstorming</li>
<li>Comprehension Strategies</li>
<li>DRTA</li>
<li>Scanning</li>
<li>SQP2RS</li>
<li>Visuals</li>
<li>Word Walls</li>
</ul></td>
</tr>
</table>

Sentence Stems and Activities Aligned to Cross-Curricular Student Expectations

(subsection c)

<table>
<tr><th colspan="3">Reading</th></tr>
<tr>
<td>4(D) continued</td>
<td>
Examples
PK-5
<ul>
<li>I know triangles have three sides.</li>
<li>I have used money to buy candy.</li>
<li>A learned perimeter is like a fence.</li>
</ul>
6-8
<ul>
<li>An example of a graph is a circle graph.</li>
<li>I think translation means to change from one language to another.</li>
<li>An equation can be represented with a model.</li>
</ul>
9-12
<ul>
<li>The pictures remind me of parallel lines.</li>
<li>I remember variables are used in equations.</li>
<li>A type of function is a linear function.</li>
</ul>
</td>
<td></td>
</tr>
<tr>
<td>4(E) read linguistically accommodated content area material with a decreasing need for linguistic accommodations as more English is learned</td>
<td>
Use of Linguistically
Accommodated Material
Teacher Questions
<ul>
<li>During lesson preparation, teachers make considerations on how to adapt reading materials based on ELLs' language proficiency levels, literacy levels in the native and target language, and educational background. When planning, consider the following questions:
<ul>
<li>How can the reading material be stated in simpler terms without diminishing the rigor of the mathematical concept?</li>
<li>What picture(s)/table/graph/manipulative/graphic organizer can be used to help students understand the reading material?</li>
<li>What irrelevant information can be deleted from the problem?</li>
</ul>
</li>
</ul>
Student Sentence Stems
<ul>
<li>___ helped me to understand/write/say ... (native language summary, native language wordlist, picture dictionary, outline, simplified English text, sentence starters, etc.)</li>
<li>The outline helped me because...</li>
<li>The problem is about...</li>
</ul>
Examples
Adapted Text
<ul>
<li>Carlos correctly answered twenty-two out of thirty-two equally weighed questions on his math test. What grade did he make on his test?</li>
</ul>
Compared to:
<ul>
<li>Carlos took a math test. Each question was worth the same amount of points. If he got 22 out of 32 problems correct, what grade did he make on his test?</li>
</ul>
</td>
<td>
<ul>
<li>Adapted Text</li>
<li>Comprehension Strategies</li>
<li>Graphic Organizers</li>
<li>Insert Method</li>
<li>Margin Notes</li>
<li>Native Language Texts</li>
<li>Outlines</li>
<li>Related Literature</li>
<li>SQP2RS</li>
<li>Taped Text</li>
</ul>
</td>
</tr>
</table>

Sentence Stems and Activities Aligned to Cross-Curricular Student Expectations

(subsection c)

<table>
<tr><th colspan="3">Reading</th></tr>
<tr>
<td>4(F) use visual and contextual support and support from peers and teachers to read grade-appropriate content area text, enhance and confirm understanding, and develop vocabulary, grasp of language structures, and background knowledge needed to comprehend increasingly challenging language</td>
<td>Using Visual/Contextual Support to Understand Text
Teacher Questions
• Based on the picture(s)/table/graph/ graphic organizer, what is the problem/text about?
• How do manipulatives (color tiles/fraction bars/algebra tiles, etc.) help you understand the reading materials?
• How could you solve this problem by using manipulatives/visualization/illustrations/diagrams?
• What is the problem situation about?

Student Sentence Stems
Reading
• The illustrations tell me this word problem/reading material is about ...
• The diagram/graph/pattern tells me the text is about ...
• The organizer tells me that I should pay attention to ...
• The organizer shows me that ___ is significant because ...
Confirming understanding
• I raise my hand when...
• I don't understand...
• I can check if I understand what I'm reading by...
• The strategy that will help me to understand this text the best is probably.... (note taking, scanning, surveying key text features, drawing, guess and check, write an equation, make a table, etc.) because...
Developing Vocabulary and Background Knowledge
• I use the word wall/wordlist while I read because...
• When I come across an unfamiliar word or phrase, I can ...
Grasp of Language Structures
• When I see ___ in a problem, it tells me....
• I noticed a lot of ____ in the problem. It probably means...

Using Teacher/Peer Support to Understand Text
Reading
• What is the problem/reading material about...?
• What does ___ mean?
• Will you read ___ for me?
• Would you please show me on the (diagram/picture/organizer /notes/etc.)....?
Confirming understanding
• It seems like ___. Is that right?
• Can you help me understand...?
• Can I please have some more information about ...?
• Where can I find out how to ...?
• Can I ask someone for help with ...?
Developing Vocabulary and Background
• Will you please explain what ___ means?
• Does ___ also mean ...?
• Why does the text have ....?
Grasp of Language Structures
• One word/expression that I saw was...
• What does the word/expression ____ mean?
• Why is there a lot of ____ in the text?</td>
<td>• Anticipation Chat
• Comprehension Strategies
• DRTA
• Graphic Organizers
• Imrov. Read Aloud
• Insert Method
• Nonlinguistic Representations
• QtA
• Question, Signal, Stem, Share, Assess
• Scanning
• SQP2RS</td>
</tr>
</table>

Sentence Stems and Activities Aligned to Cross-Curricular Student Expectations

(subsection c)

Reading		

4(G) demonstrate comprehension of increasingly complex English by participating in shared reading, retelling or summarizing material, responding to questions, and taking notes commensurate with content area and grade level needs	**Shared Reading**
	Teacher Questions
	• I will read the problem first. Who wants to re-read the problem?
	• What does ___ mean?
	• What are the most important details?
	• Why is the table/picture/graph important to understand the word problem?
	• Let's read the problem together.
	
	Student Sentence Stems
	• The problem says...
	• I will read ...
	• I'm responsible for ...
	• The table/picture/graph says...
	• The table/picture/graph is important because...
	• What does the word ___ mean?
	• My job/part/role is to...
	• Can you help me understand ...?
	• Will you please read the problem again?
	• Would you please repeat that again?
	
	Retelling/Summarizing
	Teacher Questions/Tasks
	• In your own words, what is the word problem/reading material about?
	• Prompt for elaborated student responses such as:
	  ○ Explain what ___ just said.
	  ○ Tell me more about that.
	  ○ Do you agree with ...? Why/why not?
	  ○ Why do you think...?
	
	Student Sentence Stems
	• It's about...
	• The word problem is about...
	• The problem is asking me to...
	• First.... Then.... Finally...
	• I would explain ___ to a friend by ...
	• Some ideas that could me help solve the problem include...
	
	Responding to Questions/Requests
	Teacher Questions
	• Did anyone think of this problem in a different way?
	• Why did you choose that method?
	• How does the method relate to the method ___ just explained?
	• Why did you draw ___ to represent a ___?
	• ___ please come and solve problem number ___ on the board.
	
	Student Sentence Stems
	• The answer is...
	• I think you're asking ...
	• Do you want me to ...?
	• I heard you say___, so I need to ....
		• Carousel Activity
		• Cornell Notes
		• Guess Your Corner
		• Guided Notes
		• Keep, Delete, Substitute
		• Mix and Match
		• Numbered Heads Together
		• Question, Signal, Stem, Share, Assess
		• Reciprocal Teaching
		• Poyla's Problem Solving Method
		• Story Telling
		• Structured Conversation
		• Summarization Frames

Sentence Stems and Activities Aligned to Cross-Curricular Student Expectations

(subsection c)

<table>
<tr><th colspan="3">Reading</th></tr>
<tr><td>4(G) continued</td><td>Taking Notes
Teacher Questions
• What information did you write down?
• How did you organize the information? Why?

Student Sentence Stems
• I noted ...
• I can draw a...
• I can work the problem backwards by...
• I can act it out by...
• The main ideas I wrote down were ...
• Some details I wrote down were ...
• I can organize the ideas I wrote by... (making an outline, concept map, Venn diagram, chart, etc.)</td><td></td></tr>
<tr><td>4(H) read silently with increasing ease and comprehension for longer periods</td><td>Read Silently with Increasing Comprehension
Teacher Questions
• What is the word problem asking you to do/find?
• What are some important details about the problem?
• What is the irrelevant information in the problem?

Student Sentence Stems
• I need to...
• I read about ...
• I understood/didn't understand...
• The word problem/reading material/textbook says...
• I think I need to...
• ___ is irrelevant information because...

Examples
PK-5
• I need to <u>write the number 23 in words</u>.

6-8
• I didn't understand <u>the words equally weighted in the problem</u>.
9-12
• I think I need to <u>find the zeros of the function</u>.</td><td>• Adapted Text
• Double Entry Journals
• Idea Bookmarks
• Structured Conversation</td></tr>
<tr><td>4(I) demonstrate English comprehension and expand reading skills by employing basic reading skills such as demonstrating understanding of supporting ideas and details in text and graphic sources, summarizing text, and distinguishing main ideas from details commensurate with content area needs</td><td>Supporting Ideas and Details
Graphic Sources
Teacher Questions
• What is the important information in the problem?
• What information is needed to solve the problem?
• Who...? What...? When...? How...? Which...?

Student Sentence Stems
• ___ is important.
• I highlighted/circled ___, ___, and ___.
• ___ is not needed to solve the problem.
• The illustrations tell me this problem/reading material is about ...
• This illustration/chart/diagram shows ...
• This illustration/diagram/graph/chart is significant because ...</td><td>• Comprehension Strategies
• DRTA
• Graphic Organizers
• Learning Logs
• Nonlinguistic Representation
• Numbered Heads Together
• Poyla's Problem Solving Method
• Question, Signal, Stem, Share, Assess
• QtA</td></tr>
</table>

Sentence Stems and Activities Aligned to Cross-Curricular Student Expectations

(subsection c)

<table>
<tr><th colspan="3">Reading</th></tr>
<tr><td>4(I) continued</td><td>Summarizing
Teacher Questions
• What steps will you follow to solve the word problem?
• What is the reading material/problem about?

Student Sentence Stems
• This is about ...
• I need to solve for...
• I think I need to...
• First...Second...Finally...

Distinguishing Main Ideas and Details
Teacher Questions
• What is the main idea of the word problem?
• What are the important facts that will help solve the problem?

Student Sentence Stems
• The difference between ___ and ___ is...
• The main idea of this problem is ...
• I need to ___, ___, and ___ to find the answer.
• One detail that is important to solve the problem is ...
• First I need to ___. Second I need to ___.

Examples

Word Problem
A moving company uses vans and trucks to transport furniture. The van holds up to 6 pieces of medium sized furniture. The truck holds up to 18 pieces of medium sized furniture. If 108 medium sized pieces of furniture need to be moved at one time with 8 vehicles, what system of linear equations could be used to determine the number of vans and trucks needed for the move?

Main Idea
• What system of linear equations could be used to determine the number of vans and trucks needed for the move?

Supporting Details
• Truck holds 18 pieces
• Van holds 6 pieces
• Total of 108 pieces need to be moved at one time
• 8 vehicles</td><td>• Poyla's Problem Solving Method
• Scanning
• SQP2RS
• Structured Conversation
• Summarization Frames</td></tr>
<tr><td>4(J) demonstrate English comprehension and expand reading skills by employing inferential skills such as predicting, making connections between ideas, drawing inferences and conclusions from text and graphic sources, and finding supporting text evidence commensurate with content area needs</td><td>Predicting
Teacher Questions
• Based on what you read, what do you think will happen...?
• What will happen if...?
• Do you predict the answer will be ___ or ___?

Student Sentence Stems
• I think...
• I predict ___ will happen next because...
• Based on the information in the problem/graph/table, it seems that ___ will probably....</td><td>• Comprehension Strategies
• DRTA
• Graphic Organizers
• Nonlinguistic Representations
• Scanning
• Summarization Frames
• SQP2RS</td></tr>
</table>

Sentence Stems and Activities Aligned to Cross-Curricular Student Expectations

(subsection c)

<table>
<tr><th colspan="3">Reading</th></tr>
<tr>
<td>4(J) continued</td>
<td>
Making Connections Between Ideas
Teacher Questions
• How does ___ help you understand ___?
• What is the relationship between ___ and ___?
• How else can you represent the ___?
• Which is the best model for...?
• How does ___ remind you of...?

Student Sentence Stems
• ___ reminds me of ...
• ___ is similar to ...
• ___ is different from ...
• The relationship between ___ and ___ is...
• ___ relates to what happened when ___ because...
• ___ is the result of __ because...
• If ___, then...

Drawing Inferences and Conclusions
Teacher Questions
• What can you infer/conclude from the...?
• What does infer mean?

Student Sentence Stems
• I can infer that ...
• I can assume ____ because ...
• Even though it doesn't say ___, I think ...
• Based on ___, I can conclude that ...
• From the information found in ___, I can infer that ___ because ...

Finding Supporting Text Evidence
Teacher Questions
• Based on what information can you...?
• What information supports your conclusion?
• What information supported your prediction?
• How did the graph/picture/table/manipulative help you...?

Student Sentence Stems
• The ___ helped me because...
• I think___ because ...
• ____ supports the idea that ...
• I think ___ is evidence that ...
• ___ corroborates the idea that ...
• Based on the information found in ____, I can conclude that __ because...
</td>
<td>
• Question, Signal, Stem, Share, Assess
• Structured Conversation
• Learning Logs and Journals
• QtA
• Prediction Café
• Structured Academic Controversy
</td>
</tr>
</table>

Sentence Stems and Activities Aligned to Cross-Curricular Student Expectations

(subsection c)

Reading		
4(K) demonstrate English comprehension and expand reading skills by employing analytical skills such as evaluating written information and performing critical analyses commensurate with content area and grade-level needs	**Evaluating Written Information/Performing Critical Analysis** Teacher Questions • How can you determine if your partner's solution is reasonable? • What evidence is there to support your conclusion? • Based on what information did you reach that solution? • How did you organize the information? Student Sentence Stems • A conjecture is... • A counterexample is... • I can generalize that... • The best way to represent this problem is___because... • I tried ___ but it didn't work because... • The evidence that supports the conclusion is... • The solution is/isn't reasonable because... • Your solution is reasonable/not reasonable because... • I agree/disagree with the solution because... • The conclusion to the problem is logical because... • The solution was reached by... • My partner's solution was organized/not organized clearly because...	• Book Reviews • Comprehension Strategies • Double Entry Journals • DRTA • Graphic Organizers • Learning Logs and Journals • Nonlinguistic Representations • Poyla's Problem Solving Method • QtA • Question, Signal, Stem, Share, Assess • Scanning • SQP2RS • Structured Academic Controversy • Structured Conversation • Summarization Frames

Sentence Stems and Activities Aligned to
Cross-Curricular Student Expectations
(subsection c)

Writing		
5(A) learn relationships between sounds and letters of the English language to represent sounds when writing in English	**Letter/Sound Relationship in Writing** Teacher Questions • What sound does/do the letter(s) ___ make in the word ___? • What sound does the vowel ___ make in the word ___? • Which word has the consonant blend ___? • Is ___ a long/short vowel in the word ___? • How would you write the word ___? Student Sentence Stems • ___ makes ___ sound. • ___ is pronounced ___. Therefore, it is spelled... • The letter(s) ___ make(s) the ___ sound. • The word ___ begins with the letter... • The word ___ has the long/short vowel ... • The word ___ has a silent ... • The word ___ has the consonant blend ... • The letter ___ in the word ___is long because ... • The ___ is silent in the word ___ because... • The word ___ is pronounced ___ because ... Examples PK-5 • The word eight has the silent letters gh. 6-8 • The word (teacher holds up an index card with the word Pythagorean) is pronounced Pythagorean. 9-12 • The letters PH make an "f" sound in the word graphical.	• Homophone/ Homograph Sort • Word Sorts • Word Study Books • Word Walls
5(B) write using newly acquired basic vocabulary and content-based grade-level vocabulary	**Write using New Vocabulary** Teacher Questions The following are some examples of question stems commonly used in TAKS. It is recommended for students to use terms found in these questions during writing exercises. • Which model represents___? • Which expression best represents ___? • Which expression could be used to___? • Which equation best represents the ___? • What is the equation for ___? • What is a common characteristic between ___ and ___? • Which figure has exactly ___? • If ___, what is ___? • Which ___ best represents ___? • Which table best fits the ___? • Which of these is the closest to ___? • Which ___ appears to be ___? • Which ___ describes ___? • What is the value of___? • Which ___ can be used to determine ___? • Which of the following statements is always true? • What is the effect of ___ when ___ changes?	• Choose the Words • Cloze Sentences • Dialogue Journal • Dirty Laundry • Double Entry Journals • Field Notes • Graffiti Write • Learning Logs and Journals • Read, Write, Pair, Share • Roundtable • Self-assessment of Levels of Word Knowledge • Think, Pair, Share • Word Sort • Word Wall • Ticket Out

Sentence Stems and Activities Aligned to Cross-Curricular Student Expectations

(subsection c)

<table>
<tr><th colspan="3">Writing</th></tr>
<tr><td>5(B) continued</td><td>Student Sentence Stems<ul><li>___ best represents.___.</li><li>___ could be used to___.</li><li>___ is for ___.</li><li>A common characteristic between ___ and ___ is...</li><li>___ has exactly ___.</li><li>If ___, then ___.</li><li>___ best fits ___.</li><li>___ is the closest to ___.</li><li>___ appears to be ___.</li><li>___ describes ___.</li><li>The solution is...</li><li>The effect is...</li><li>The value of ___ is...</li><li>___ can be used to determine ___.</li><li>The ___ that is always true is...</li><li>___ will make the equation true.</li><li>If the pattern continues, it will take...</li><li>At this rate, it will take...</li></ul>Examples<ul><li>For examples of high frequency words, content vocabulary and sight words go to student expectations 3(B), 3(D), and 4(C).</li></ul></td><td></td></tr>
<tr><td>5(C) spell familiar English words with increasing accuracy, and employ English spelling patterns and rules with increasing accuracy as more English is acquired</td><td>English Spelling
Patterns and Rules
Teacher Questions<ul><li>How is the word ___ spelled?</li><li>Does the word ___ start with ___ or ___?</li><li>What does the word ___ start/end with?</li><li>Is the word spelled with the vowel ___ or ___?</li><li>Did you write the word ___ with the vowel/letter ___ or ___?</li><li>What are words students commonly misspell?</li></ul>Student Sentence Stems<ul><li>___ is spelled ...</li><li>___ begins/ends with the letter ...</li><li>I spelled the word ___ with a ___.</li><li>In this set of words I notice ...</li><li>These words are all similar because ...</li><li>The spelling rule that applies to this word is ___ because ...</li><li>This word is spelled correctly/incorrectly because ...</li><li>I can check my spelling by ...</li><li>Is this the correct spelling for...?</li><li>Is ___ spelled with a ___ or ___?</li><li>How do you spell...?</li><li>Will you please check the spelling of the word ____?</li></ul></td><td><ul><li>Homophone/ Homograph Sort</li><li>Peer Editing</li><li>Personal Spelling Guide</li><li>Word Analysis</li><li>Word Sorts</li><li>Word Walls</li></ul></td></tr>
</table>

Sentence Stems and Activities Aligned to Cross-Curricular Student Expectations

(subsection c)

Writing		
5(D) edit writing for standard grammar and usage, including subject-verb agreement, pronoun agreement, and appropriate verb tenses commensurate with grade-level expectations as more English is acquired	**Grammar and Usage** Teacher Questions • What are some common editing symbols? • How can you use editing symbols to help your classmate correct grammatical mistakes? • Did you use the correct verb agreement, pronoun agreement or verb tense in your sentence? How do you know? • Which verb would you use to...? • When do you use __ instead of ___? Student Sentence Stems • The subject ___ agrees/disagrees with the verb___ because... • The pronoun ___ agrees/disagrees with ___ because... • The present/past/future/conditional tense is appropriate/inappropriate in this sentence because ... Example PK-5 • Coin add to 50. VS. Coins add up to 50 cents. • Daniela said he taked half of the base ten blocks. VS. Daniela said she took half of the base ten blocks. 6-8 • The graph are all circle graphs. VS. The graphs are all circle graphs. 9-12 • The radiuses is proportion. VS. The radii are proportional.	• Contextualized Grammar Instruction • Daily Oral Language • Oral Scaffolding • Peer Editing • Reciprocal Teaching • Sentence Mark Up • Sentence Sort
5(E) employ increasingly complex grammatical structures in content area writing commensurate with grade-level expectations, such as: (i) using correct verbs, tenses, and pronouns/ antecedents; (ii)using possessive case (apostrophe s) correctly; and (iii)using negatives and contractions correctly	Teacher Questions • Are students writing and speaking using correct verbs? • Are students using double negatives? • When should students use NON as opposed to NOT? • Are students using correct pronouns? Student Sentence Stems **Using Correct Verb Tenses** • The ___ is ___. • These ___ are ____. • I predict the ___ will... • I concluded that... • The answer is/isn't... • I do/don't agree with...because... • My prediction was not correct because... • In math, NON means...	• Contextualized Grammar Instruction • Daily Oral Language • Oral Scaffolding • Peer Editing • Reciprocal Teaching • Sentence Mark Up • Sentence Sort

Sentence Stems and Activities Aligned to Cross-Curricular Student Expectations

(subsection c)

<table>
<tr><th colspan="3">Writing</th></tr>
<tr><td rowspan="2">5(E) continued</td><td>Using Possessive Case/Contractions Correctly

• The graph's ___ shows…
• The number's ___ is…
• My partners' ideas were…
• The table's ___ has…
• The function's ___ is…
• The answer can't be ___ because…
• The ___ isn't ____.
• ___ isn't a characteristic of…
• ___ doesn't best represent…
• ___ isn't reasonable because…

Using Negatives
• ___ is a non-example.
• ___ is a non-example because…
• ___ is NOT an example of…
• A counterexample is…
• The solution is not…</td><td rowspan="2"></td></tr>
<tr><td>Examples
PK-5
• The bear's color is blue.
• A curve isn't a characteristic of a quadrilateral.
• A box is a non-example of a 2 dimensional figure.
6-8
• The circle's area is shaded blue and its circumference is highlighted in yellow.
• The integer's sign is negative.
• The solution isn't the square root of 5.
9-12
• The function's slope is 3/2.
• The plane's points aren't collinear.</td></tr>
<tr><td>5(F) write using a variety of grade-appropriate sentence lengths, patterns, and connecting words to combine phrases, clauses, and sentences in increasingly accurate ways as more English is acquired</td><td>Speak using a variety of Structures

Teacher Questions
• In writing, explain…
  ○ What are the attributes of…?
  ○ How would you order ___ from ___ to ___?
  ○ What will happen if…?
  ○ What are the similarities between ___ and ___?
  ○ Predict what the next ___ will be.
  ○ What can you infer from the ___?
  ○ What can you conclude from ___?

Student Sentence Stems
Description
• A ___ has ___.
• A ___ has ___, ___, and ___.
• A ___ is ___, ___, and ___.
• Additionally ___ has …
• ___ is an example of…
• ___ is an example of…. because …</td><td>• Dialogue Journal
• Double Entry Journals
• Draw & Write
• Field Notes
• Free Write
• Genre Analysis /Imitation
• Hand Motions for Connecting Words
• Letters/Editorials
• Learning Logs and Journals
• Perspective-Based Writing
• Read, Write, Pair, Share
• Summarization Frames</td></tr>
</table>

Sentence Stems and Activities Aligned to Cross-Curricular Student Expectations

(subsection c)

<table>
<tr><th colspan="3">Writing</th></tr>
<tr><td>5(E) continued</td><td>Sequence
• First___. Second___.
• First ___ and then...
• If numbers ___ are ordered from ___ to ___, the first number would be...
Cause and Effect
• The ___ is ___.
• ___ because ___.
• The cause is ___. The effect is ___.
• ___ was caused by ___.
• If ___, then ___.
• When ___then...
• The independent variable is ___, and the dependent variable is___.
Comparison
• A ___ has ___.
• A ___ has ___ but a ___ has ___.
• ___ and ___ both have...
• ___ is the same as ___.
• ___ differs from ___ in that...
• Although ___ has ___, ___ has ___.
• ___on the other hand has...
Predictions
• The ___ will have...
• The ___ will be...
• I predict ___ will...
• I predict ___ will ___ because...
• The next pattern will be...
• Due to ___, I think ___ will happen.
• Consequently, I think...
Inferences
• I can infer that...
• I know ___ because...
• My conjecture is...
• From the ___, I can infer...
Conclusion
• All ___ are ___.
• ___ are ___.
• I concluded...
• I can conclude that...
• If ___, then ___. Therefore...</td><td></td></tr>
<tr><td></td><td>Examples</td><td></td></tr>
<tr><td></td><td>PK-5
Description
• A pentagon has sides.
• A pentagon has 5 sides, 5 vertices, and 5 angles.
Sequence
• First you add. Second you subtract.
• First you add and then subtract.</td><td></td></tr>
</table>

Sentence Stems and Activities Aligned to Cross-Curricular Student Expectations

(subsection c)

Writing		
5(F) continued	**Examples** 6-8 Cause and Effect • The pattern is getting bigger. • The pattern is getting bigger because you add three color tiles each time. Comparison • A bar graph has bars. • A bar graph has bars but a circle graph has sectors. Prediction • The marble will be red. • I predict the next marble will be red because there are more red than blue marbles. 9-12 Inferences • I infer that the graph will not cross the x-axis. • My conjecture is that the pattern represents the function y=2x+1.. Conclusion • All functions are relations.	
5(G) narrate, describe, and explain with increasing specificity and detail to fulfill content area writing needs as more English is acquired	**Narrate, Describe, and Explain with Increasing Detail** Teacher Questions • How would you describe ...? • In your own words, explain... • Why did ___ happen? • What else can you say about...? • Will you please restate what ___ said? • Explain how you got to that solution. Student Sentence Stems • ___ is ___. • This is a ... • This is a... and it has/is___ and ___. • The ___ is... • The ___ is ___ because... • ___ best represents ___. • ___ best represents ___ because... • ___ is about... • The most important attribute is... • ___ is the most important attribute because... • It's important to remember... • Initially...., then..., ultimately... • First..., then..., finally... **Examples** PK-5 • This is a square. • This is a square and it has 4 sides and 4 angles. 6-8 • The equation is y=2x. • The equation is y=2x because the pattern grows by two tiles each time. 9-12 • The most important attribute is rate of change. • The most important attribute is the slope because it is always increasing at a constant rate.	• Free Write • Learning Logs and Journals • Dialogue Journal • Field Notes • Double Entry Journals • Draw & Write • Perspective-Based Writing • Unit Study for ELLs

Language Proficiency Level Descriptors

(subsection d)

§74.4. English language Proficiency Standards

http://www.tea.state.tx.us/rules/tac/chapter074/ch074a.html.

(d) Proficiency level descriptors.

(1) Listening, Kindergarten-Grade 12. ELLs may be at the beginning, intermediate, advanced, or advanced high stage of English language acquisition in listening. The following proficiency level descriptors for listening are sufficient to describe the overall English language proficiency levels of ELLs in this language domain in order to linguistically accommodate their instruction.

(A) Beginning. Beginning ELLs have little or no ability to understand spoken English in academic and social settings. These students:

(i) struggle to understand simple conversations and simple discussions even when the topics are familiar and the speaker uses linguistic supports such as visuals, slower speech and other verbal cues, and gestures;

(ii) struggle to identify and distinguish individual words and phrases during social and instructional interactions that have not been intentionally modified for ELLs; and

(iii) may not seek clarification in English when failing to comprehend the English they hear; frequently remain silent, watching others for cues.

(B) Intermediate. Intermediate ELLs have the ability to understand simple, high-frequency spoken English used in routine academic and social settings. These students:

(i) usually understand simple or routine directions, as well as short, simple conversations and short, simple discussions on familiar topics; when topics are unfamiliar, require extensive linguistic supports and adaptations such as visuals, slower speech and other verbal cues, simplified language, gestures, and preteaching to preview or build topic-related vocabulary;

(ii) often identify and distinguish key words and phrases necessary to understand the general meaning during social and basic instructional interactions that have not been intentionally modified for ELLs; and

(iii) have the ability to seek clarification in English when failing to comprehend the English they hear by requiring/requesting the speaker to repeat, slow down, or rephrase speech.

(C) Advanced. Advanced ELLs have the ability to understand, with second language acquisition support, grade-appropriate spoken English used in academic and social settings. These students:

(i) usually understand longer, more elaborated directions, conversations, and discussions on familiar and some unfamiliar topics, but sometimes need processing time and sometimes depend on visuals, verbal cues, and gestures to support understanding;

(ii) understand most main points, most important details, and some implicit information during social and basic instructional interactions that have not been intentionally modified for ELLs; and

(iii) occasionally require/request the speaker to repeat, slow down, or rephrase to clarify the meaning of the English they hear.

(D) Advanced high. Advanced high ELLs have the ability to understand, with minimal second language acquisition support, grade-appropriate spoken English used in academic and social settings. These students:

(i) understand longer, elaborated directions, conversations, and discussions on familiar and unfamiliar topics with occasional need for processing time and with little dependence on visuals, verbal cues, and gestures; some exceptions when complex academic or highly specialized language is used;

(ii) understand main points, important details, and implicit information at a level nearly comparable to native English-speaking peers during social and instructional interactions; and

(iii) rarely require/request the speaker to repeat, slow down, or rephrase to clarify the meaning of the English they hear.

(2) Speaking, Kindergarten-Grade 12. ELLs may be at the beginning, intermediate, advanced, or advanced high stage of English language acquisition in speaking. The following proficiency level descriptors for speaking are sufficient to describe the overall English language proficiency levels of ELLs in this language domain in order to linguistically accommodate their instruction.

(A) Beginning. Beginning ELLs have little or no ability to speak English in academic and social settings. These students:

(i) mainly speak using single words and short phrases consisting of recently practiced, memorized, or highly familiar material to get immediate needs met; may be hesitant to speak and often give up in their attempts to communicate;

(ii) speak using a very limited bank of high-frequency, high-need, concrete vocabulary, including key words and expressions needed for basic communication in academic and social contexts;

(iii) lack the knowledge of English grammar necessary to connect ideas and speak in sentences; can sometimes produce sentences using recently practiced, memorized, or highly familiar material;

(iv) exhibit second language acquisition errors that may hinder overall communication, particularly when trying to convey information beyond memorized, practiced, or highly familiar material; and

(v) typically use pronunciation that significantly inhibits communication.

(B) Intermediate. Intermediate ELLs have the ability to speak in a simple manner using English commonly heard in routine academic and social settings. These students:

(i) are able to express simple, original messages, speak using sentences, and participate in short conversations and classroom interactions; may hesitate frequently and for long periods to think about how to communicate desired meaning;

(ii) speak simply using basic vocabulary needed in everyday social interactions and routine academic contexts; rarely have vocabulary to speak in detail;

(iii) exhibit an emerging awareness of English grammar and speak using mostly simple sentence structures and simple tenses; are most comfortable speaking in present tense;

(iv) exhibit second language acquisition errors that may hinder overall communication when trying to use complex or less familiar English; and

(v) use pronunciation that can usually be understood by people accustomed to interacting with ELLs.

(C) Advanced. Advanced ELLs have the ability to speak using grade-appropriate English, with second language acquisition support, in academic and social settings. These students:

(i) are able to participate comfortably in most conversations and academic discussions on familiar topics, with some pauses to restate, repeat, or search for words and phrases to clarify meaning;

(ii) discuss familiar academic topics using content-based terms and common abstract vocabulary; can usually speak in some detail on familiar topics;

(iii) have a grasp of basic grammar features, including a basic ability to narrate and describe in present, past, and future tenses; have an emerging ability to use complex sentences and complex grammar features;

(iv) make errors that interfere somewhat with communication when using complex grammar structures, long sentences, and less familiar words and expressions; and

(v) may mispronounce words, but use pronunciation that can usually be understood by people not accustomed to interacting with ELLs.

(D) Advanced high. Advanced high ELLs have the ability to speak using grade-appropriate English, with minimal second language acquisition support, in academic and social settings. These students:

(i) are able to participate in extended discussions on a variety of social and grade-appropriate academic topics with only occasional disruptions, hesitations, or pauses;

(ii) communicate effectively using abstract and content-based vocabulary during classroom instructional tasks, with some exceptions when low-frequency or academically demanding vocabulary is needed; use many of the same idioms and colloquialisms as their native English-speaking peers;

(iii) can use English grammar structures and complex sentences to narrate and describe at a level nearly comparable to native English-speaking peers;

(iv) make few second language acquisition errors that interfere with overall communication; and

(v) may mispronounce words, but rarely use pronunciation that interferes with overall communication.

(3) Reading, Kindergarten-Grade 1. ELLs in Kindergarten and Grade 1 may be at the beginning, intermediate, advanced, or advanced high stage of English language acquisition in reading. The following proficiency level descriptors for reading are sufficient to describe the overall English

language proficiency levels of ELLs in this language domain in order to linguistically accommodate their instruction and should take into account developmental stages of emergent readers.

(A) Beginning. Beginning ELLs have little or no ability to use the English language to build foundational reading skills. These students:

(i) derive little or no meaning from grade-appropriate stories read aloud in English, unless the stories are:

(I) read in short "chunks;"

(II) controlled to include the little English they know such as language that is high frequency, concrete, and recently practiced; and

(III) accompanied by ample visual supports such as illustrations, gestures, pantomime, and objects and by linguistic supports such as careful enunciation and slower speech;

(ii) begin to recognize and understand environmental print in English such as signs, labeled items, names of peers, and logos; and

(iii) have difficulty decoding most grade-appropriate English text because they:

(I) understand the meaning of very few words in English; and

(II) struggle significantly with sounds in spoken English words and with sound-symbol relationships due to differences between their primary language and English.

(B) Intermediate. Intermediate ELLs have a limited ability to use the English language to build foundational reading skills. These students:

(i) demonstrate limited comprehension (key words and general meaning) of grade-appropriate stories read aloud in English, unless the stories include:

(I) predictable story lines;

(II) highly familiar topics;

(III) primarily high-frequency, concrete vocabulary;

(IV) short, simple sentences; and

(V) visual and linguistic supports;

(ii) regularly recognize and understand common environmental print in English such as signs, labeled items, names of peers, logos; and

(iii) have difficulty decoding grade-appropriate English text because they:

(I) understand the meaning of only those English words they hear frequently; and

(II) struggle with some sounds in English words and some sound-symbol relationships due to differences between their primary language and English.

(C) Advanced. Advanced ELLs have the ability to use the English language, with second language acquisition support, to build foundational reading skills. These students:

(i) demonstrate comprehension of most main points and most supporting ideas in grade-appropriate stories read aloud in English, although they may still depend on visual and linguistic supports to gain or confirm meaning;

(ii) recognize some basic English vocabulary and high-frequency words in isolated print; and

(iii) with second language acquisition support, are able to decode most grade-appropriate English text because they:

(I) understand the meaning of most grade-appropriate English words; and

(II) have little difficulty with English sounds and sound-symbol relationships that result from differences between their primary language and English.

(D) Advanced high. Advanced high ELLs have the ability to use the English language, with minimal second language acquisition support, to build foundational reading skills. These students:

(i) demonstrate, with minimal second language acquisition support and at a level nearly comparable to native English-speaking peers, comprehension of main points and supporting ideas (explicit and implicit) in grade-appropriate stories read aloud in English;

(ii) with some exceptions, recognize sight vocabulary and high-frequency words to a degree nearly comparable to that of native English-speaking peers; and

(iii) with minimal second language acquisition support, have an ability to decode and understand grade-appropriate English text at a level nearly comparable to native English-speaking peers.

(4) Reading, Grades 2-12. ELLs in Grades 2-12 may be at the beginning, intermediate, advanced, or advanced high stage of English language acquisition in reading. The following proficiency level descriptors for reading are sufficient to describe the overall English language proficiency levels of ELLs in this language domain in order to linguistically accommodate their instruction.

(A) Beginning. Beginning ELLs have little or no ability to read and understand English used in academic and social contexts. These students:

(i) read and understand the very limited recently practiced, memorized, or highly familiar English they have learned; vocabulary predominantly includes:

(I) environmental print;

(II) some very high-frequency words; and

(III) concrete words that can be represented by pictures;

(ii) read slowly, word by word;

(iii) have a very limited sense of English language structures;

(iv) comprehend predominantly isolated familiar words and phrases; comprehend some sentences in highly routine contexts or recently practiced, highly familiar text;

(v) are highly dependent on visuals and prior knowledge to derive meaning from text in English; and

(vi) are able to apply reading comprehension skills in English only when reading texts written for this level.

(B) Intermediate. Intermediate ELLs have the ability to read and understand simple, high-frequency English used in routine academic and social contexts. These students:

(i) read and understand English vocabulary on a somewhat wider range of topics and with increased depth; vocabulary predominantly includes:

(I) everyday oral language;

(II) literal meanings of common words;

(III) routine academic language and terms; and

(IV) commonly used abstract language such as terms used to describe basic feelings;

(ii) often read slowly and in short phrases; may re-read to clarify meaning;

(iii) have a growing understanding of basic, routinely used English language structures;

(iv) understand simple sentences in short, connected texts, but are dependent on visual cues, topic familiarity, prior knowledge, pretaught topic-related vocabulary, story predictability, and teacher/peer assistance to sustain comprehension;

(v) struggle to independently read and understand grade-level texts; and

(vi) are able to apply basic and some higher-order comprehension skills when reading texts that are linguistically accommodated and/or simplified for this level.

(C) Advanced. Advanced ELLs have the ability to read and understand, with second language acquisition support, grade-appropriate English used in academic and social contexts. These students:

(i) read and understand, with second language acquisition support, a variety of grade-appropriate English vocabulary used in social and academic contexts:

(I) with second language acquisition support, read and understand grade-appropriate concrete and abstract vocabulary, but have difficulty with less commonly encountered words;

(II) demonstrate an emerging ability to understand words and phrases beyond their literal meaning; and

(III) understand multiple meanings of commonly used words;

(ii) read longer phrases and simple sentences from familiar text with appropriate rate and speed;

(iii) are developing skill in using their growing familiarity with English language structures to construct meaning of grade-appropriate text; and

(iv) are able to apply basic and higher-order comprehension skills when reading grade-appropriate text, but are still occasionally dependent on visuals, teacher/peer assistance, and other linguistically accommodated text features to determine or clarify meaning, particularly with unfamiliar topics.

(D) Advanced high. Advanced high ELLs have the ability to read and understand, with minimal second language acquisition support, grade-appropriate English used in academic and social contexts. These students:

(i) read and understand vocabulary at a level nearly comparable to that of their native English-speaking peers, with some exceptions when low-frequency or specialized vocabulary is used;

(ii) generally read grade-appropriate, familiar text with appropriate rate, speed, intonation, and expression;

(iii) are able to, at a level nearly comparable to native English-speaking peers, use their familiarity with English language structures to construct meaning of grade-appropriate text; and

(iv) are able to apply, with minimal second language acquisition support and at a level nearly comparable to native English-speaking peers, basic and higher-order comprehension skills when reading grade-appropriate text.

(5) Writing, Kindergarten-Grade 1. ELLs in Kindergarten and Grade 1 may be at the beginning, intermediate, advanced, or advanced high stage of English language acquisition in writing. The following proficiency level descriptors for writing are sufficient to describe the overall English language proficiency levels of ELLs in this language domain in order to linguistically accommodate their instruction and should take into account developmental stages of emergent writers.

(A) Beginning. Beginning ELLs have little or no ability to use the English language to build foundational writing skills. These students:

(i) are unable to use English to explain self-generated writing such as stories they have created or other personal expressions, including emergent forms of writing (pictures, letter-like forms, mock words, scribbling, etc.);

(ii) know too little English to participate meaningfully in grade-appropriate shared writing activities using the English language;

(iii) cannot express themselves meaningfully in self-generated, connected written text in English beyond the level of high-frequency, concrete words, phrases, or short sentences that have been recently practiced and/or memorized; and

(iv) may demonstrate little or no awareness of English print conventions.

(B) Intermediate. Intermediate ELLs have a limited ability to use the English language to build foundational writing skills. These students:

(i) know enough English to explain briefly and simply self-generated writing, including emergent forms of writing, as long as the topic is highly familiar and concrete and requires very high-frequency English;

(ii) can participate meaningfully in grade-appropriate shared writing activities using the English language only when the writing topic is highly familiar and concrete and requires very high-frequency English;

(iii) express themselves meaningfully in self-generated, connected written text in English when their writing is limited to short sentences featuring simple, concrete English used frequently in class; and

(iv) frequently exhibit features of their primary language when writing in English such as primary language words, spelling patterns, word order, and literal translating.

(C) Advanced. Advanced ELLs have the ability to use the English language to build, with second language acquisition support, foundational writing skills. These students:

(i) use predominantly grade-appropriate English to explain, in some detail, most self-generated writing, including emergent forms of writing;

(ii) can participate meaningfully, with second language acquisition support, in most grade-appropriate shared writing activities using the English language;

(iii) although second language acquisition support is needed, have an emerging ability to express themselves in self-generated, connected written text in English in a grade-appropriate manner; and

(iv) occasionally exhibit second language acquisition errors when writing in English.

(D) Advanced high. Advanced high ELLs have the ability to use the English language to build, with minimal second language acquisition support, foundational writing skills. These students:

(i) use English at a level of complexity and detail nearly comparable to that of native English-speaking peers when explaining self-generated writing, including emergent forms of writing;

(ii) can participate meaningfully in most grade-appropriate shared writing activities using the English language; and

(iii) although minimal second language acquisition support may be needed, express themselves in self-generated, connected written text in English in a manner nearly comparable to their native English-speaking peers.

(6) Writing, Grades 2-12. ELLs in Grades 2-12 may be at the beginning, intermediate, advanced, or advanced high stage of English language acquisition in writing. The following proficiency level descriptors for writing are sufficient to describe the overall English language proficiency levels of ELLs in this language domain in order to linguistically accommodate their instruction.

(A) Beginning. Beginning ELLs lack the English vocabulary and grasp of English language structures necessary to address grade-appropriate writing tasks meaningfully. These students:

(i) have little or no ability to use the English language to express ideas in writing and engage meaningfully in grade-appropriate writing assignments in content area instruction;

(ii) lack the English necessary to develop or demonstrate elements of grade-appropriate writing such as focus and coherence, conventions, organization, voice, and development of ideas in English; and

(iii) exhibit writing features typical at this level, including:

(I) ability to label, list, and copy;

(II) high-frequency words/phrases and short, simple sentences (or even short paragraphs) based primarily on recently practiced, memorized, or highly familiar material; this type of writing may be quite accurate;

(III) present tense used primarily; and

(IV) frequent primary language features (spelling patterns, word order, literal translations, and words from the student's primary language) and other errors associated with second language acquisition may significantly hinder or prevent understanding, even for individuals accustomed to the writing of ELLs.

(B) Intermediate. Intermediate ELLs have enough English vocabulary and enough grasp of English language structures to address grade-appropriate writing tasks in a limited way. These students:

(i) have a limited ability to use the English language to express ideas in writing and engage meaningfully in grade-appropriate writing assignments in content area instruction;

(ii) are limited in their ability to develop or demonstrate elements of grade-appropriate writing in English; communicate best when topics are highly familiar and concrete, and require simple, high-frequency English; and

(iii) exhibit writing features typical at this level, including:

(I) simple, original messages consisting of short, simple sentences; frequent inaccuracies occur when creating or taking risks beyond familiar English;

(II) high-frequency vocabulary; academic writing often has an oral tone;

(III) loosely connected text with limited use of cohesive devices or repetitive use, which may cause gaps in meaning;

(IV) repetition of ideas due to lack of vocabulary and language structures;

(V) present tense used most accurately; simple future and past tenses, if attempted, are used inconsistently or with frequent inaccuracies;

(VI) undetailed descriptions, explanations, and narrations; difficulty expressing abstract ideas;

(VII) primary language features and errors associated with second language acquisition may be frequent; and

(VIII) some writing may be understood only by individuals accustomed to the writing of ELLs; parts of the writing may be hard to understand even for individuals accustomed to ELL writing.

(C) Advanced. Advanced ELLs have enough English vocabulary and command of English language structures to address grade-appropriate writing tasks, although second language acquisition support is needed. These students:

(i) are able to use the English language, with second language acquisition support, to express ideas in writing and engage meaningfully in grade-appropriate writing assignments in content area instruction;

(ii) know enough English to be able to develop or demonstrate elements of grade-appropriate writing in English, although second language acquisition support is particularly needed when topics are abstract, academically challenging, or unfamiliar; and

(iii) exhibit writing features typical at this level, including:

(I) grasp of basic verbs, tenses, grammar features, and sentence patterns; partial grasp of more complex verbs, tenses, grammar features, and sentence patterns;

(II) emerging grade-appropriate vocabulary; academic writing has a more academic tone;

(III) use of a variety of common cohesive devices, although some redundancy may occur;

(IV) narrations, explanations, and descriptions developed in some detail with emerging clarity; quality or quantity declines when abstract ideas are expressed, academic demands are high, or low-frequency vocabulary is required;

(V) occasional second language acquisition errors; and

(VI) communications are usually understood by individuals not accustomed to the writing of ELLs.

(D) Advanced high. Advanced high ELLs have acquired the English vocabulary and command of English language structures necessary to address grade-appropriate writing tasks with minimal second language acquisition support. These students:

(i) are able to use the English language, with minimal second language acquisition support, to express ideas in writing and engage meaningfully in grade-appropriate writing assignments in content area instruction;

(ii) know enough English to be able to develop or demonstrate, with minimal second language acquisition support, elements of grade-appropriate writing in English; and

(iii) exhibit writing features typical at this level, including:

(I) nearly comparable to writing of native English-speaking peers in clarity and precision with regard to English vocabulary and language structures, with occasional exceptions when writing about academically complex ideas, abstract ideas, or topics requiring low-frequency vocabulary;

(II) occasional difficulty with naturalness of phrasing and expression; and

(III) errors associated with second language acquisition are minor and usually limited to low-frequency words and structures; errors rarely interfere with communication.

(e) Effective date. The provisions of this section supersede the ESL standards specified in Chapter 128 of this title (relating to Texas Essential Knowledge and Skills for Spanish Language Arts and English as a Second Language) upon the effective date of this section.

Source: http://www.tea.state.tx.us/rules/tac/chapter074/ch074a.html

ELPS Linguistic Accommodation by Proficiency Level Self-Assessment

Rate the current level of awareness of the English Language Proficiency standards at your district or campus.

A: Always
M: Mostly
S: Sometimes
N: Never

Indicator	A	M	S	N	Comments/Questions
I am aware of the level of language proficiency of the English learners I teach.					
I am aware of specific instructional strategies to support ELLs at various levels of English language proficiency.					
I am aware of specific English language district and classroom resources that enhance comprehension for ELLs at various levels of proficiency.					
I am aware of specific native language district and classroom resources that enhance comprehension for ELLs at various levels of proficiency.					
I differentiate instruction to meet students' needs at various levels of language proficiency.					
I provide a variety of resources for English learners at various levels of proficiency.					

Summaries of ELPS Proficiency Level Descriptors*

Please refer to actual proficiency level descriptors to plan instruction.

Level	Listening (*d1: k-12*) *The student comprehends...*	Speaking (*d2: k-12*) *The student speaks...*	Reading (*d4: 2-12*) *The student reads...*	Writing (*d6: 2-12*) *The student writes...*
Beginning (A)	1A(i) few **simple conversations** with linguistic support 1A(ii) **modified conversation** 1A(iii) few words, **does not seek clarification**, watches others for cues	2A(i) using **single words and short phrases** with practiced material; tends to give up on attempts 2A(ii) using **limited bank of key vocabulary** 2A(iii) with **recently practiced familiar material** 2A(iv) with frequent **errors** that hinder communication 2A(v) with **pronunciation that inhibits communication**	4A(i) little except recently practiced terms, **environmental print**, high frequency words, **concrete words represented by pictures** 4A(ii) **slowly, word by word** 4A(iii) with very limited sense of English structure 4A(iv) with comprehension of **practiced, familiar text** 4A(v) with need for **visuals and prior knowledge** 4A(vi) modified and **adapted text**	6A(i) with **little ability to use English** 64A(ii) **without focus** and coherence, conventions, organization, voice 6A(iil) labels, lists, and copies of printed text and **high-frequency words/phrases**, short and simple, practiced sentences primarily in **present tense with frequent errors** that hinder or prevent understanding
Intermediate (B)	1B(i) unfamiliar language with **linguistic supports** and adaptations 1B(ii) unmodified conversation with **key words** and phrases 1B(iii) with **requests for clarification** by asking speaker to repeat, slow down, or rephrase speech	2B(i) **with simple messages** and hesitation to think about meaning 2B(ii) using **basic vocabulary** 2B(iii) with **simple sentence structures** and present tense 2B(iv) with **errors** that inhibit unfamiliar communication 2B(v) with **pronunciation generally understood** by those familiar with English language learners	4B(i) **wider range of topics**: and everyday academic language 4B(ii) **slowly** and **rereads** 4B(iii) basic language structures 4B(iv) simple sentences **with visual cues, pretaught vocabulary** and **interaction** 4B(v) **grade-level texts** with difficulty 4B(vi) at high level with **linguistic accommodation**	6B(i) with **limited ability to use English** in content area writing 6B(ii) best on **topics that are highly familiar** with simple English 6B(iii) with **simple oral tone in messages**, high-frequency vocabulary, loosely connected text, repetition of ideas, **mostly in the present tense**, undetailed descriptions, and **frequent errors**
Advanced (*C*)	1C(i) with some processing time, **visuals, verbal cues, and gestures; for unfamiliar conversations** 1C(ii) most unmodified interaction 1C(iii) with occasional **requests** for the speaker to slow down, repeat, rephrase, and **clarify meaning**	2C(i) in conversations with some **pauses to restate, repeat, and clarify** 2C(ii) using **content-based and abstract terms** on familiar topics 2C(iii) with **past, present, and future** 2C(iv) using **complex sentences** and grammar with some errors 2C(v) with **pronunciation usually understood** by most	4C(i) abstract grade appropriate text 4C(ii) **longer phrases and familiar sentences** appropriately 4C(iii) while developing the ability to construct meaning from text 4C(iv) at **high comprehension** level with **linguistic support** for unfamiliar topics and to clarify meaning	6C(i) grade appropriate **ideas with second language support** 6C(ii) with extra need for second language **support when topics are technical and abstract** 6C(iii) with a grasp of basic English usage and some understanding of complex usage with **emerging grade-appropriate vocabulary** and a more academic tone
Advanced High (*D*)	1D(i) longer **discussions on unfamiliar topics** 1D(ii) spoken information nearly **comparable to native speaker** 1D(iii) with few requests for speaker to slow down, repeat, or rephrase	2D(i) in **extended discussions** with few pauses 2D(ii) using **abstract content-based vocabulary** except low frequency terms; using idioms 2D(iii) with grammar **nearly comparable to native speaker** 2D(iv) with **few errors** blocking communication 2D(v) **occasional mispronunciation**	4D(i) **nearly comparable to native speakers** 4D(ii) **grade appropriate familiar text** appropriately 4D(iii) while constructing meaning at near native ability level 4D(iv) with high level comprehension with **minimal linguistic support**	6D(i) grade appropriate **content area ideas with little need for** linguistic **support** 6D(ii) develop and demonstrate **grade appropriate** writing 6D (iii) nearly **comparable to native speakers** with clarity and precision, with **occasional difficulties** with naturalness of language.

***These summaries are not appropriate to use in formally identifying student proficiency levels for TELPAS. TELPAS assessment and training materials are provided by the Texas Education Agency Student Assessment Division**: http://www.tea.state.tx.us/index3.aspx?id=3300&menu_id3=793*

Linguistic Accommodations for each Proficiency Level*

Sequence of Language Development	Communicating and Scaffolding Instruction			
	Listening *Teachers...*	**Speaking** *Teachers...*	**Reading** *Teachers...*	**Writing** *Teachers...*
Beginning Students (A)	• Allow use of same language peer and **native language support** • Expect student to struggle to understand simple conversations • Use **gestures and movement** and other linguistic support to communicate language and expectations	• Provide short **sentence stems** and single words for practice before conversations • **Allow some nonparticipation** in simple conversations • Provide **word bank** of key vocabulary • Model **pronunciation of social and academic language**	• Organize reading in **chunks** • Practice **high frequency, concrete terms** • Use **visual and linguistic supports** • Explain classroom **environmental print** • Use adapted text	• Allow **drawing and use of native language** to express concepts • Allow student to use high frequency recently memorized, and **short, simple, sentences** • Provide **short, simple sentence stems** with present tense and high frequency vocabulary
Intermediate (B)	• Provide **visuals, slower speech, verbal cues, simplified language** • **Preteach vocabulary** before discussions and lectures • **Teach phrases** for student to request speakers repeat, slow down, or rephrase speech	• Allow extra **processing time** • Provide **sentence stems** with simple sentence structures and tenses • Model and provide practice in **pronunciation of academic terms**	• Allow wide range of reading • Allow grade-level comprehension and **analysis of tasks** including **drawing** and use of **native language** and peer collaboration • Provide high level of **visual and linguistic supports** with adapted text and **pretaught vocabulary**	• Allow **drawing and use of native language** to express academic concepts • Allow writing on **familiar, concrete topics** • **Avoid assessment of language errors** in content area writing • Provide simple **sentence stems and scaffolded writing assignments**
Advanced (C)	• Allow some **processing time, visuals, verbal cues, and gestures** for unfamiliar conversations • Provide opportunities for student to **request clarification**, repetition and rephrasing	• **Allow extra time** after pauses • Provide **sentence stems** with past, present, future, and **complex grammar,** and vocabulary with **content-based and abstract terms**	• Allow abstract grade-level reading comprehension and analysis with **peer support** • Provide **visual and linguistic supports** including **adapted text** for unfamiliar topics	• Provide **grade-level appropriate writing tasks** • Allow abstract and technical writing with linguistic support including teacher **modeling and student interaction** • Provide complex **sentence stems** for **scaffolded writing assignments**
Advanced High (D)	• Allow some **extra time** when academic material is complex and unfamiliar • Provide **visuals, verbal cues, and gestures** when material is complex and unfamiliar	• Opportunities for extended **discussions** • Provide **sentence stems with** past, present, future, and **complex grammar** and vocabulary with **content-based and abstract terms**	• Allow abstract grade-level reading • Provide minimal **visual and linguistic supports** • Allow grade level comprehension and **analysis tasks** with **peer collaboration**	• Provide complex **grade-level appropriate writing tasks** • Allow abstract and technical writing with **minimal linguistic support** • Use **genre analysis** to identify and use features of advanced English writing

**Guidelines at specific proficiency levels may be beneficial for students at all levels of proficiency depending on the context of instructional delivery, materials, and students' background knowledge.*

Differentiating by Language Level
Instructional Planning Guide

Advanced/Advanced High	Intermediate	Beginners
• Visuals for academic vocabulary and concepts	• Visuals for academic vocabulary and concepts	• Visuals for classroom vocabulary and academic concepts
• Grade-level text	• Adapted grade level text	• Native language and adapted grade level text
• Complex sentence stems	• Sentence stems	• Short, simple sentence stems
• Preteaching low-frequency academic vocabulary	• Preteaching academic Vocabulary	• Preteaching social and academic vocabulary
• Peer interaction	• Peer Interaction	• Peer interaction (same language peer as needed)
• Verbal scaffolding as needed	• Verbal scaffolding	• Extensive verbal scaffolding
• Grade level writing tasks	• Adapted writing tasks with scaffolding	• Adapted writing tasks with drawing and scaffolding
• Gestures for memorization of academic concepts	• Gestures for memorization of academic concepts	• Gestures (basic and academic concepts)
• Modeling	• Modeling	• Modeling
• Graphic organizers	• Graphic organizers	• Graphic organizers
• Manipulatives	• Manipulatives	• Manipulatives
		• Preteaching functional language (stems for social interaction)
		• Pronunciation of social/academic language
		• Slower, simplified speech
		• Instruction in high frequency concrete social vocabulary
		• Use of native language for key concepts
		• Verbal cues
		• Chunking use of information in print
		• Word bank

Differentiating by Language Level
Instructional Planning Template

Grade Level/Topic: Key Vocabulary & Concepts:	Content Objective: Language Objective:	
Tasks and Accommodations for Advanced/Advanced High	Accommodation to Support Intermediate Students	Accommodations to Support Beginners

This page is intentionally left blank.

Guide to Terms and Activities

Accountable Conversation Questions – Place the following poster in your room:

> **What to say instead of "I Don't Know"**
>
> - May I please have some more information?
> - May I have some time to think?
> - Would you please repeat the question?
> - Where could I find information about that?
>
> *Please speak in complete sentences.*

Model for the students how to use the questions when they are unsure what to say when called on by a teacher (Seidlitz & Perryman, 2008). Explain that they are called on for a response, they can either respond, or ask for help and then respond. Newcomer English learners should not be pressured to speak in front of the class if they have not yet begun to show early production levels of speech proficiency. Students should be encouraged, but not forced to speak when in the silent period of language development (Krashen, 1982).

Adapted Text – Techniques for making the content presented in texts available to students who are not able to fully comprehend the level of academic language including: graphic organizers, outlines, highlighted text, taped text, margin notes, native language texts, native language glossaries and word lists (Echevarria, Vogt & Short, 2008)

Advance Organizers – Information given to students prior to reading or instruction that helps them organize the information they will encounter during instruction (Mayer, 2003). Advance organizers should involve both activating prior knowledge and organizing new information. Examples include: graphic organizers, anticipation guides, KWL, guided notes, etc.

Anticipation Chat – Prior to instruction, a teacher facilitates a conversation between students about the content to be learned. The teacher opens the discussion by having the students make inferences about what they are going to learn based on their prior knowledge and experiences and limited information about the new concepts (Zwiers, 2008).

Anticipation Guides: A structured series of statements given to students before instruction. Students choose to agree or disagree with the statements either individually or in groups. After instruction, students revisit the statements and discuss whether they agree or disagree with them again after having learned about the topic. (Head, M. H. & Readence, J. 1986).

Backwards Book Walk: Students scan a non-fiction text briefly looking at headings, illustrations, captions, key words, and other text features. After the scan, students discuss what they believe they will learn from the text. (Echevarria & Vogt, 2008)

Book Reviews: After being immersed in the book review genre, English learners write short reviews which can then be published so that others can have access to them. (Samway, K., 2006)

Brick and Mortar Cards: Students are given five "brick" cards with academic vocabulary (content area terms) and are instructed to organize them however they think makes sense. Afterward, they have to link the cards together using language. They write the language they are using on "mortar" cards that they then use to tie the concepts together. Students may need lists of sentence terms and connecting words to facilitate the process. (Zwiers, 2008)

CALLA Approach: An approach to teaching English learners which involves the explicit teaching of language learning strategies and academic content and language skills through scaffolding and active engaged learning and language use. (Chamot, A. & O'Malley, J., 1994)

CCAP (Communicative Cognitive Approach to Pronunciation): A five step process for assisting English learners in improving pronunciation. (Celce-Murcia, M., Brinton. D. & Goodwin. J, 1996 as cited in Flores M., 1998)

- Description and analysis of the pronunciation feature.
- Listening/Discrimination activities (see segmental/supra segmental practice below).
- Controlled practice and feedback
- Guided practice and feedback
- Communicative practice

Canned Questions: Students are given a series of question stems ranging from the lowest to the highest level of Blooms taxonomy so that they can participate in discussions about a topic (Echevarria & Vogt, 2008). For example:

- "What is...?"
- "How do..."
- "What would be a better approach to...?"
- "How do you know that...?"

Carousel Activity: This activity encourages interaction among students as they answer questions posted around the room. Groups are assigned to a station and are given a specified time to answer the station's question. Groups rotate around the room until they have answered all questions. (Comments from CRISS, 1996)

Chat Room: This writing activity allows students to use informal and formal English to describe terms/concepts. Each student is given a term/concept and a paper outline of a computer screen. On the computer screen, students describe the term/concept by writing a text message. Students are then paired and switch computer screens with their partner. Partners need to rewrite the text message using formal English.

Choose the Words: Students select words from a word wall or word list to use in a conversation or in writing.

Chunking Input: Breaking up material into smaller units for easier comprehension. Visual and auditory information can be chunked so that students have time to discuss new information, pay attention to details, and create schema for organizing new information.

Cloze Sentences: Fill in the blank sentences used to help students process academic text. (Taylor, 1953; Gibbons, 2002)

Compare, Contrast, Analogy & Metaphor Frames: Sentence frames used to help students organize schema for new words (Marzano, 2001 & Hill, J. & Flynn, K. 2006)

. For example:

- Compare: ___ is similar to ___ in that both....
- Contrast: ___ is different from ___ in that ...
- Analogy: ___ is to ___ as ___ is to ___
- Metaphor: I think ___ is like/is... because...

Comprehension Strategies: Strategies used by proficient readers to understand what they read. These strategies are used in different kinds of text, can be taught, and when they are taught, students are likely to use them. Strategies include: prediction, self questioning, monitoring, determining importance, and summarizing. (Echevarria, Vogt, & Short, 2008; Dole, Duffy, Roehler & Pearson, 1991; Baker, 2004)

Concept Attainment: A strategy based on the theories of Jerome Bruner in which the teacher gives students examples and non examples of a concept and has students categorize them. Over time students develop conceptual categories at increasing levels of depth and understanding. (Boulware, B.J., & Crow, M., 2008; Bruner, J., 1967)

Concept Definition Map: A visual organizer that enables students to process a term. (Echevarria, Vogt, & Short, 2008). Four questions are asked:

- What is the term?
- What is it?
- What is it like?
- What are some examples?

Concept Mapping: A technique for making a visual diagram of the relationship between concepts. Concept maps begin with a single concept which is written in a square or circle. New concepts are listed and connected with lines and shapes creating a web showing the relationship between the ideas. (Novak, J.D., 1995)

Conga Line: Students form two lines facing one another. Students in each row share ideas, review concepts, or ask one another questions. After the first discussion, one row moves and the other remains stationary so that each student now has a new partner. (Echevarria & Vogt, 2008)

Content-Specific Stems: Sentence stems using content specific vocabulary. For example, instead of a general stem such as "In my opinion..." a content specific stem would be "In my opinion the Declaration of Independence is significant because..."

Contextualized Grammar Instruction: Teaching grammar in mini-lessons that apply to specific, meaningful tasks students will perform. The purpose of the grammar instruction is to enable students to communicate verbally or in writing more effectively. (Weaver, 1996)

Cornell Notes: A method of note taking in which a paper is divided into two columns. In one large column students take traditional notes in modified outline form. In the other column, students write key vocabulary terms and questions. (Paulk, Walter, 2000).

Creating Analogies: Generating comparisons using the frame: ____ is to ____ as ___ is to ____. (Marzano, R., Pickering, D., & Pollock, J, 2001)

Creating Words: This vocabulary game provides students the opportunity to review key vocabulary by representing words in creative ways. A student selects a word and rolls a cube which has options on each face such as model it, draw it, act it out, write it, talk about it, etc. Based on the outcome of the rolled cube, the student represents the word and classmates guess the name of the term.

Daily Oral Language: A strategy for teaching English usage which involves five minute mini lessons where students view a list of sentences that have incorrect English usage. Students learn correct usage by correcting the mistakes in the sentences. (Vail, N. & Papenfuss, J., 1993).

Dialog Journal: A journal that is exchanged between the student and teacher or between two or more students. The journal focuses on academic topics and the language used by the teacher and student should be content focused and academic. (Samway, K., 2006)

Direct Teaching of Affixes: Lessons on prefixes and suffixes to build knowledge of English word structure. (White, Sowell, & Yanagihara, 1989)

Direct Teaching of Cognates: Lessons on words that sound the same in the primary language and the target language. For a list of Spanish and English cognates see: http://www.colorincolorado.org/pdfs/articles/cognates.pdf . Students must be careful of false cognates, words that sound the same in the primary and target language, but do not have the same meaning. For a list of false Spanish/English cognates see: http://www.platiquemos-letstalk.com/Extras/Articles/FalseCognates/FalseCongnatesMain.htm

Direct Teaching of Roots: Teaching students Greek and Latin roots that form the base of many words in English. A partial list of roots can be found here: https://www.msu.edu/~defores1/gre/roots/gre_rts_afx2.htm

Directionality Sort: Students are given copies of texts in various languages in groups. Each group must sort the texts based on perceived directionality. Is the text written from top to bottom then left to right? Is the text right to left, then top to bottom? For newspapers to use showing letters and characters used in a variety of languages see: www.newoxxo.com

Dirty Laundry: This vocabulary activity helps students extend their knowledge of newly acquired terms. Students are given one vocabulary word or content concept and an outline of a paper t-shirt, pant, dress or skirt. On one side of the t-shirt, students write a message of their assigned word without using the word. On the reverse side of the t-shirt, students draw a picture of the word/concept. The aim of this activity is to have other students in the class guess what word each t-shirt describes. T-shirts can then be displayed on walls or hung using clothes pins. (Created by Cristina Ferrari, Brownsville ISD)

Discovery Learning: An inquiry-based approach to instruction in which teachers create problems and dilemmas through which students construct knowledge and representations of knowledge. Ideas, hypotheses, and explanations continue to be revised as learning takes place. (Bruner, J.S. 1967). The approach has been criticized by some (Marzano, 2001; Kirschner, P. A., Sweller, J. & Clark, R. E. (2006) for teaching skills to novices who don't have adequate background and language to be able to use a discovery approach to efficiently learn new content. Teachers of English learners must be careful to preteach content area functional language and set goals and objectives for the lesson when teaching English learners using a discovery approach.

Discussion Starter Cards: Small cards containing sentence starters for students to use when beginning an academic conversation, or seeking ways to extend a conversation. For example: In my opinion..., I think..., another possibility is ... etc. (Thornburry, 2005)

Double Entry Journals: A two column journal used for reflective writing about texts. In one column students write words, phrases, or ideas that they found interesting or significant. In the other column, students write the reasons they found them significant or ways they could use them in their own writing. (Samway, K., 2006)

Draw & Write: Allowing English learners to express their knowledge of academic content using both drawing and writing. Students may use their native language to express ideas but are encouraged to express new concepts using English. (Adapted from: Samway, K., 2006)

DRTA: Directed Reading-Thinking Activity. This activity involves the teacher stopping regularly before and during reading to have the students make predictions and justify their predictions. Questions might be: What do you think is going to happen? Why do you think that will happen next? Is there another possibility? What made you think that? (Echevarria, Vogt, & Short, 2008)

Experiments/Labs: A form of discovery learning in science where students directly encounter the scientific process: Making an observation, forming a hypothesis, testing the hypothesis, and coming to a conclusion. Teachers of ELLs need to make sure to preteach necessary content and functional vocabulary to enable full participation of English learners.

Expert/Novice: A simulation involving two students. One student takes on the role of an expert and the other a novice in a particular situation. The expert responds to questions asked by the novice. The procedure can be used for lower level cognitive activities such as having students introduce one another to classroom procedures, and higher level activities such as explaining content area concepts at greater degrees of depth. The procedure can also be used to model the difference between formal and informal English, with the expert speaking formally and the novice informally. (Seidlitz & Perrryman, 2008)

Field Notes: Students take notes and write in a journal and write reflections about what they are learning and experiencing. Field journals should be content focused yet can contain both social and academic language as well as drawing. (Samway, K., 2006)

Flash Card Review: Students make flash cards, preferably including images with explanations of the meanings of words. Students study, play games, and sort the flash cards in various ways.

Fluency Workshop: Students have three opportunities to talk and listen to another student talk about the same topic. They alternate roles back and forth from listening to speaking. When listening, they may ask questions, but cannot contribute an opinion on what the speaker has said. After the activity students reflect on their level of fluency in the first discussion and the third discussion. (Maurice, K., 1983).

Fold the Line: Students line up chronologically based on a predetermined characteristic such as height, age, number of pets, etc. The line then folds in half upon itself providing each student to have a partner. Students are then asked to formulate a response/answer to a task or question. Depending on the task/question, students use formal or informal English to share responses with their partner. (Kagan, 1992)

Formal/Informal Pairs: The teacher provides strips of paper with pairs of statements written in formal English and in informal English. The papers are distributed to the students and students have to find the student who has their match. Students can also be given sets of all the pairs in small groups and sort them into two stacks.

Four Corners Vocabulary: A way of processing vocabulary with a paper or note card divided into four sections: the term, a definition, a sentence, and an illustration. (Developed by D. Short, Center for Applied Linguistics. Described in: Echevarria & Vogt, 2008)

Framed Oral Recap: An oral review involving two students using sentence starters. Students are given stems such as: "Today I realized...," "Now I know....," and "The most significant thing I learned was ...." They pair up with a partner to discuss what they have learned in a lesson or unit (Adapted from Zwiers, 2008).

Free Write: Students write nonstop about a topic for five to ten minutes. The goal is to keep writing, even if they can't think of ideas. They may write "I don't know what to write" if they are unable to think of new ideas during the process. English learners can use sketching and write in the native language during the process although they can be encouraged to write in English. (Elbow, P. 1998)

Genre Analysis/Imitation: Students read high quality selections from a genre of literature. They note particular words, phrases and ideas they found interesting or effective and record those in a journal. Students then use their notes and observations as a resource when writing in that genre. (Adapted from Samway, K., 2006)

Graffiti Write: This instructional technique is used to access prior knowledge or review key content concepts/vocabulary. Students are grouped in sets of 3-5. Each group is provided with a chart paper that contains the concept or key term in the center of the paper. The sheet is divided into the total number of students in the group. Students are given 2-3 minutes to write linguistic and nonlinguistic representations of what they know or learned about the concept/term. (Think Literacy: Cross-Curricular Approaches, 2003)

Graphic Organizers: A way of developing a learner's schema by organizing information visually. Examples include the T-Chart, Venn diagram, Concept Map, Concept Web, Timeline, etc. Graphic organizers are a form of nonlinguistic representation that can help students process and retain new information. (Marzano, R., Pickering, D. & Pollock., J., 2001)

Group Response with a White Board: Students write responses to questions on white boards using dry erase markers. These can be made from card stock slipped into report covers, or with shower board cut into squares able to fit on student's desks. White boards are a form of active response signal that research has shown to be highly effective in improving achievement for struggling learners.

Guess Your Corner: This activity serves as a way review or to assess student understanding of key content concepts. Begin by posting four previously introduced terms/content concepts around the room. Each student is then given a characteristic, attribute, picture, synonym, etc. of one of the four terms/content concepts. The responsibility of each student is to guess their correct corner.

Guided Notes: Teacher prepared notes used as a scaffold to help students take notes during a lecture and learn note taking skills. For examples of guided note formats see: http://www.studygs.net/guidednotes.htm

Hand Motions for Connecting Words: Gestures representing transition/signal words that students use to visually model the function of connecting words in a sentence. For example students might bring their hands together for terms like: also, including, as well as, etc. For terms such as excluding, neither, without, no longer, etc., students could bring their hands together. Students can come up with their own signals for various categories including: comparing, contrasting, cause and effect, sequence, description, and emphasis. (Adapted from: Zwiers, 2008)

Hi-Lo Readers: Readers published on a variety of reading levels while having the same content focus and objectives. For example National Geographic Explorer Books found here: http://new.ngsp.com/Products/SocialStudies/nbspnbspNationalGeographicExplorerBooks/tabid/586/Default.aspx

And http://www.kidbiz3000.com/

Homophone/Homograph Sort: Teacher prepares cards with words that sound or are written the same but are spelled differently such as *know* and *no* or *rose* (a flower) and *rose* (past tense of rise). The teacher asks the students to group the words that sound the same together and then explain the meanings of each.

IEPT: Inter-Ethnolingusitic Peer Tutoring: A research based method for increasing fluency in English learners by pairing them up with fluent English speakers. Tasks are highly structured and fluent English speakers are trained to promote more extensive interaction with English learners (Johnson. D. 1995).

Idea Bookmarks: Students take reflective notes on bookmark size pieces of paper. The bookmarks include quotes, observations, and words that strike the reader as interesting or effective. The bookmarks can be divided into boxes as quotes are added with page numbers written in each box (Samway, K., 2006).

Imrov Read Aloud: Students act out a story silently while the teacher or another student reads aloud. Each student has a role and has to discover how to act out the story while it is being read. Afterward, students discuss how each student represented their part during the improv. (Zwiers, 2008)

Insert Method: Students read text with a partner and mark the texts with the following coding system: a *check* to show a concept or fact already known, a *question mark* to show a concept that is confusing, an *exclamation mark* to show something new or surprising, or a *plus* to show an idea or concept that is new. (Echevarria & Vogt, 2008)

Inside/Outside Circle: A way of facilitating student conversations. Students form two concentric circles facing one another, an inside circle and an outside circle. Students can then participate in short, guided discussion or review with their partner. After the discussion, the outside circle rotates one person to the right while the inside circle remains still. All students now have a new partner to speak with. (Kagan, 1990)

Instructional Conversation: A way of engaging students in conversation about literature through open ended dialog between the teacher and students in small groups. Instructional conversations are open ended, have few "known answer" questions, and promote complex language and expression. (Goldenberg, C., 1992)

Instructional Scaffolding: A model of teaching where students achieve increasing levels of independence following the pattern: teach, model, practice, and apply. (Echevarria, Vogt & Short, 2008)

Interactive Reading Logs: Reading journals where students write reflections to texts read silently. These logs can be exchanged with other students or with the teacher who can write questions or responses to what students have written. These logs are ideal components of an SSR program.

Interview Grids: A grid used to get students to record other student's responses to various questions. Students wander around the room and search for their partners who will respond to their questions. (Zwiers, 2008)

Keep, Delete, Substitute, Select: A strategy for summarizing developed by Brown, Campoine, and Day (1981) discussed in *Classroom Instruction That Works* (Marzano. R, Pickering D., & Pollock J., 2001) Students keep important information, delete unnecessary and redundant material, substitute general terms for specific terms (e.g. birds for robins, crows, etc.), and select or invent a topic sentence. For ELLs, Hill and Flynn (2006) recommend using gestures to represent each phase of the process and clearly explain the difference with high frequency and low frequency terms.

KWL: A prereading strategy used to access prior knowledge and set up new learning experiences (Ogle, 1986). The teacher creates a chart where students respond to three questions. The first two are discussed prior to reading or the learning experience, and the third is discussed afterward.

Learning Logs and Journals: Students record observations and questions about what they are learning in a particular content area. The teacher can provide general or specific sentence starters to help students begin their reflections. (Samway, K., 2006)

Letters/Editorials: Students write letters and editorials from their own point of view or from the point of view of a character in a novel, person from history, or a physical object (sun, atom, frog, etc.) Teachers of ELLs should remember to scaffold the writing process by providing sentence frames, graphic organizers, wordlists, and other supports. Newcomers may use the Draw/Write method discussed above.

List Stressed Words: Students take a written paragraph and highlight words that would be stressed, focusing on stressing content English words such as nouns, verbs, adverbs over process words such as articles, prepositions, linking-verbs/modals and auxiliaries.

List/Group/Label: Students are given a list of words or students brainstorm a list of words. They then sort the words into similar piles and then create labels for each pile. This can be done by topic (planets, stars, scientific laws etc.) or by word type (those beginning with a particular letter, those with a particular suffix, and those in a particular tense) (Taba, Hilda, 1967)

Literature Circles: Activity through which students form small groups similar to "book clubs" to discuss literature. Roles include: discussion facilitators, passage pickers, illustrators, connectors, summarizers, vocabulary enrichers, travel tracers, investigators, and figurative language finders. ELLs will need to be supported with sentence starters, wordlists, and adapted text as necessary depending on language level. (Schlick, N. & Johnson, N., 1999). For support in starting literature circles see: http://www.litcircles.org/ .

Margin Notes: A way of adapting text. Teachers, students, or volunteers write key terms, translations of key terms or short native language summaries, text clarifications, or hints for understanding in the margins of a text book. (Echevarria, Vogt & Short, 2008)

Math Sorts: A sorting activity in which students classify numbers, equations, geometric shapes, etc. based on given categories. For example, students are given 20 systems of linear equations cards and their responsibility is to determine whether each system belongs under the category of parallel lines, perpendicular lines or neither.

Mix and Match: This activity encourages students to interact with classmates and practice their formal and informal English. Each student is provided with a card that has information matching another student's card. When the teacher says mix, students stand and walk around the room until the teacher says match. Students find their match by using the sentence stem, "I have ____. Who has ____?"

Multiple Representations Card Game: This activity is a variation of the Spoons Card Game. Depending on the number of cards, students play in groups of 3-5. The objective of the game is to be the first player to get all the representations of a particular math concept.

Multiple Representations Graphic Organizer (MRGO): An instructional tool used to illustrate an algebraic situation in multiple representations including a picture, graph, table, equation, and verbal description. (Echevarria, Short, & Vogt, 2009)

Native Language Texts: Native language translations, chapter summaries, wordlists, glossaries, or related literature that can be used to understand texts used in content area classes. Many text book companies include Spanish language resources with the adoption.

Nonlinguistic Representations: Nonverbal means of representing knowledge including illustrations, graphic organizers, physical models, and kinesthetic activities (Marzano, R., Pickering, D., & Pollock, J., 2001). Hill, J. and Flynn, K. (2006) advocate integrating Total Physical Response (Asher J., 1967) as a means of integrating nonlinguistic representations because of its unique way of engaging learners especially those in the early stages of language development.

Note Taking Strategies: Strategies for organizing information presented in lectures and in texts. English learners at the early stages of language development benefit from guided notes (see above), native language wordlists, summaries, and opportunities to clarify concepts with peers. Strategies include informal outlines, concept webbing, Cornell Note taking, and combination notes. Research seems to indicate that students should write more rather than less when taking notes (Marzano R., Pickering D., and Pollock J., 2001). ELLs in pre-production phases can respond to teacher notes through gesture. Those in early production and speech emergent phases can communicate about information in teacher prepared notes using teacher provided sentence frames. (Hill, J. & Flynn, K., 2006)

Numbered Heads Together: A strategy for having all students in a group share with the whole class over time. Each student in a group is assigned a number (1, 2, 3 and 4). When asking a question the teacher will ask all the ones to speak first, and then open up the discussion to the rest of the class. For the next question the teacher will ask the two's to speak, then the threes, and finally the fours. The teacher can also randomize which number will speak in which order. When doing numbered heads together with English learners, teachers should provide sentence starters for the students. (Kagan, 1992).

Oral Scaffolding: The process of:

- explicitly teaching academic language
- modeling academic language
- providing opportunities in structured ways for students to use language orally

- write using the language they have already seen modeled and have used. (Adapted from Gibbons, 2002)

Order It Up Math Puzzle: A number sentence or equation is written on a sentence strip. The sentence strip is cut into individual pieces and placed in an envelope. Students work in pairs to determine the correct order of each piece to come up with the original number sentence/equation. (Created by Amy King, Independent Consultant)

Outlines: Traditional note taking method involving roman numerals, Arabic numerals, upper and lowercase letters.

Pairs View: A strategy for keeping students engaged and focused while they process viewed material at a deeper level. When watching a video clip or movie, each pair is assigned a role. For example, one partner might be responsible for identifying key dates while another is listing important people and their actions. (Kagan, S., 1992).

Partner Reading: A strategy for processing text where two students read a text. Each can alternate a paragraph while the other summarizes or one can read and the other student summarize and ask questions. (Johnson, D., 1995)

Peer Editing: Students review one another's work using a rubric. Research shows that English learners benefit from peer editing when trained on specific strategies for participating in peer response to writing. (Berg, C., 1999)

Personal Dictionary: Students choose words from the word wall, other wordlists, and words encountered in texts to record on note cards or in a note book which become a personal dictionary. Students are encouraged to use drawing, reflection, and their native language when explaining the meaning of terms. (Adapted from Echevarria, Vogt, & Short, 2008)

Personal Spelling Guide: Students record correct spellings of misspelled words on note cards. As the number of cards grows, students sort the words based on characteristics of the words. Students should generate the categories for example, students may develop lists like: contractions, big words, words with "ie" or "ei", words that are hard to say, words I will never use. Encourage students to look for patterns in the spellings of the words. Students can select a number of words to review and have a partner quiz them orally over their self-selected words.

Perspective-Based Writing: Writing from an assigned point of view using specific academic language. For example, students in a social studies class could write from the perspective of Martin Luther King writing a letter explaining his participation in the Montgomery bus boycott to a fellow pastor. Students should be given specific words and phrases to integrate into the writing assignment. Students can also write from the point of view of inanimate objects such as rocks, water, molecules, etc. and describe processes from an imaginative perspective as if they were that object. In addition, students can take on the role of an expert within a field: math, science, social studies, literature and use the language of the discipline to write about a particular topic. Genre studies can be particularly helpful as a way of preparing students for perspective-based writing activities. (Seidlitz & Perryman, 2008).

Posted Phrases and Stems: Sentence frames posted in clearly visible locations in the classroom to enable students to have easy access to functional language during a task. For example, during a lab the teacher might post the stems: How do I record...., Can you help me (gather, mix, measure, identify, list...., Can you explain what you mean by ...? Frames should be posted in English but can be written in the native language as well.

Prediction Café- A way of having students participate in mini discussions about what will happen or what students will learn about in a text. Pick out important headings, quotes, or captions from a text (about eight quotes for a class of 24). . They may only speak with one student at a time. Some students may have the same card. Either way, they will discuss with that student what they think the text is about or what they think will happen in the text. Students should be given frames to facilitate the development of academic language during the activity such as: *__makes me think that.., I believe ___ because..., etc.).* (Zwiers, J., 2008)

Pretest with a partner: Students are given a pretest in pairs. Students take turns reading the questions. After each question they try to come to consensus before recording an answer. (Echevarria, J. & Vogt, M., 2008)

Polya's Problem Solving Method: A four step model for solving word problems.

- Step 1: Understanding the problem
- Step 2: Devising a plan
- Step 3: Carrying out the plan
- Step 4: Check

QtA (Question the Author): A strategy for deepening the level of thinking about literature (Beck. I. & McKeown, M., Hamilton, R., & Kugan. L., 1997) Instead of staying within the world of the text, the teacher prompts the students to question the author. For example:

- What do you think the author is trying to say?
- Why do you think the author chose that word or phrase?
- Would you have chosen a different word or phrase?

Question Answer Relationship (QAR): A way of teaching students to analyze the nature of questions they are asked about a text. Questions are divided into four categories (Echevarria J., & Vogt M., 2008)

- Right there (found in the text)
- Think and Search (require thinking about relationships between ideas in the text)
- Author and Me (require me to form an inference about the text)
- On My Own (requires me to reflect on my own experience and knowledge)

Question, Signal, Stem, Share, Assess: A strategy to get students to use new academic language during student-student interactions. The teacher asks a question and then asks students to show a signal when they are ready to respond to the question with a particular sentence stem provided by the teacher. When all students are ready to share, they share their answers. Students are then assessed either through random calling on individual students after the conversation or through writing assignments that follow the conversation (Seidlitz, J., & Perryman B., 2008).

Radio Talk Show: Students create a radio talk show about a particular topic. This can be a good opportunity for students to practice using academic language as they take on the role of an expert. It can also provide an opportunity for students to identify the distinctions between formal and informal use of English as they take on different roles. (Wilhelm, J., 2002)

R.A.F.T.: A social studies writing strategy that enables students to write from various points of view (Fisher D. & Frey N., 2007). The letters stand for Role (the perspective the students take), Audience (the individuals the author is addressing), Format (type of writing that will take place), Topic (the subject).

Read, Write, Pair, Share: A strategy for getting students to share their writing and ideas during interactions. Students read a text, write their thoughts on it using a sentence starter, pair up with another student and share their writing. Students can also be given suggestions on how to comment on one another's writing. (Fisher, D., & Frey, N., 2007).

Reader/Writer/Speaker Response Triads: A way of processing text in cooperative groups. Students form groups of three. One student will read the text aloud; one will write the group's reactions or responses to questions about the text, a third will report the answers to the group. After reporting to the group, the students switch roles. (Echevarria J., & Vogt M., 2008)

Recasting: Repeating an English learner's incorrect statement or question correctly without changing the meaning in a low risk environment where the learner feels comfortable during the interaction. Recasts have been shown to have a positive impact on second language acquisition (Leeman, J., 2003).

Reciprocal Teaching: A student-student interaction involving collaboration in creating meaning from texts (Palincsar & Brown, 1985) Hill and Flynn (2006) suggest adapting reciprocal teaching for use among English learner's by providing vocabulary, modeling language use, and using pictorial representation during the discussion. Reciprocal teaching involves a student leader that guides the class through stages: Summarizing, Question Generating, Clarifying, and Predicting.

Related Literature: Connecting and supporting texts used in content areas. These texts can be fiction or nonfiction, in the native language or the target language. (Echevarria J., & Vogt M., Short. D., 2008)

ReQuest: A variation of reciprocal teaching (see above). The teacher asks questions using particular stems following a period of silent reading. After another period of silent reading, the teacher provides the stems for the students and has them ask the questions over the text. (Manzo, A., 1969: as cited in Fisher D. & Frey N., 2007)

Retelling: Students retell a narrative text in their own words or summarize an expository text in their own words.

Roundtable: This is a cooperative learning technique in which small groups are given a paper with a category, term or task. The paper goes around the table and each group member is responsible for writing a characteristic/synonym/ step of task. (Kagan, 1992)

Same Scene Twice: Students perform a skit involving individuals discussing a topic. The first time through, the individuals are novices who use informal language to discuss the topic. The second time through they are experts who discuss the topic using correct academic terminology and academic English (adapted from Wilhelm, J., 2002).

Scanning: Students scan through a text backwards looking for unfamiliar terms. The teacher then provides quick brief definitions for the terms giving the students only the meaning of the word as it appears in context. Marzano, Pickering and Pollock (2001) state that "even superficial instruction on words greatly enhances the probability that student will learn the words from context when they encounter them in their reading and that, "the effects of vocabulary instruction are even more powerful when the words selected are those that students most likely will encounter when they learn new content."

Segmental Practice: Listening/Discrimination activities that help learners listen for and practice pronouncing individual combinations of syllables. There are several ways to engage in segmental practice. Tongue twisters and comparisons with native language pronunciations can help English learners practice English pronunciation. The activity "syllable, storm, say" involves students brainstorming syllables that begin with a particular sound for example: pat pen pal pas pon pem, etc. Long and short vowel sounds can be used as well as diphthongs. Students then practice in partners pronouncing the terms. (Celce-Murcia, M., Brinton, D. & Goodwin, J., 1996).

Self-assessment of Levels of Word Knowledge: Students rank their knowledge of new words from the word wall and other word lists using total response signals (see below) or sentence starters. Responses range from no familiarity with the word to understanding a word well and being able to explain it to others. (Diamond & Gutlohn, 2006: as cited in Echevarria, Vogt, Short, 2008)

Sentence Mark Up: Method of using colored pencils to mark texts to indicate cause and effect, opposing thoughts, connecting words, and other features of a sentence to understand the relationship between clauses. (Zwiers, J., 2008)

Sentence Sort: Sorting various sentences based on characteristics. The teacher provides the sentences and students sort them. This can be done with an open sort where students create the categories or a closed sort where the teacher creates the categories. It can also be done by taking a paragraph from a text book or piece of literature the students are going to read and using sentences from the text. Possible categories include:

- Description sentences
- Complex sentences
- Simple sentences
- Sentences connecting ideas
- Sentences comparing ideas
- Sentences opposing ideas
- Sentences with correct usage
- Sentences with incorrect usage
- Sentences in formal English
- Sentences in informal English

Sentence Stems: Incomplete sentences provided for students to help scaffold the development of specific language structures and to facilitate entry into conversation and writing. For example "In my opinion..." and "One characteristic of annelids is...

Signal Words: Words that determine a text pattern such as generalization, cause and effect, process, sequence, etc. A sample of signal words can be found at: http://www.nifl.gov/readingprofiles/Signal_Words.pdf

Six Step Vocabulary Process: Research based process developed by Marzano (2004) that teachers can use to develop academic vocabulary. The steps are: Teacher provides a description. Students restate the explanation in their own words. Students create a nonlinguistic representation of the term. Students periodically do activities that help them add to their knowledge of vocabulary terms. Periodically students are asked to discuss the terms with each other. Periodically, students are involved in games that allow them to play with the terms.

Sound Scripting: A way for students to mark text showing pauses and stress. Students use a writing program to write a paragraph and then enter a paragraph break to show pauses and capital and bold letters to show word stress. (Powell, M., 1996)

SQP2RS (Squeepers): A classroom reading strategy that trains students to use cognitive metacognitive strategies to process nonfiction text. The following steps are involved (Echevarria, Vogt, Short, 2008):

- Survey: students scan the visuals, headings, and other text features.
- Question: students write what questions they might find answers to
- Predict: student write predictions about what they will learn
- Read: students read the text
- Respond: revisit your questions and think through how you respond to how you read
- Summarize: Students restate key concepts either individually or on groups

SSR Program (Sustained, Silent Reading): A program used by schools to encourage students to read silently to develop literacy where students read whatever they wish for fifteen to twenty minutes during a school day. Pilgreen (2000) discusses eight features of high quality SSR programs: Access to books, book appeal, conducive environment, encouragement to read, non-accountability, distributed reading time, staff training, and follow up activities. (Pilgreen, 2000).

Story Telling: Students retell narratives in their own language.

Structured Academic Controversy A way of structuring classroom discussion to promote deep thinking and the taking of multiple perspectives. Johnson & Johnson (1995) outline five steps.

- Organizing Information And Deriving Conclusions
- Presenting And Advocating Positions
- Uncertainty Created By Being Challenged By Opposing Views
- Epistemic Curiosity And Perspective Taking
- Reconceptualizing, Synthesizing, and Integrating

Structured Conversation: Students-student interaction where the language and content are planned. Students are given sentence frames to begin the conversation and specific questions and sentence starters to extend the conversation.

Summarization Frames: A way of structuring summaries of content area text. The frames involve specific questions that help students summarize different kinds of texts Marzano (2001 p. 27-42) and Hill & Flynn (2006) discuss seven frames:

- narrative frame
- topic restriction frame
- illustration frame
- definition frame
- argumentation frame
- problem solution frame
- conversation frame

Suprasegmental Practice: Pronunciation practice involving units groups of syllables. Some techniques include: sound scripting (see above), recasting (see above), a pronunciation portfolio, and content/function word comparisons. (Wennerstrom, A., 1993).

Systematic Phonics Instruction: teaching sound-spelling relationships and how to use those relationships to read. The national literacy panel (Francis, D.J., Lesaux, N.K., & August, D.L., 2006) reported that instruction in phonemic awareness, phonics, and fluency had "clear benefits for language minority students."

Taped Text: Recordings of text used as a way of adapting text for English Learners. (Echevarria, Vogt, & Short, 2008)

Think Alouds: A way for teachers to scaffold cognitive and metacognitive thinking by saying aloud the thinking involved in solving problems and making decisions. (Bauman, Russel & Jones, 1992)

Think, Pair, Share: A method of student-student interaction. The teacher asks a question and then provides wait time. The students then find a partner and share their answers. Afterward, selected students share their thoughts with the whole class. (Lyman, 1981)

Ticket Out: A short reflection written at the end of a lesson. Teachers can use tickets out as an opportunity for students to reflect on what they have learned and use new vocabulary by specifying specific words and phrases for students to use.

Tiered Questions: Asking varying the type of questions students are asked based on their level of language development. (Hill & Flynn, 2006)

Tiered Response Stems: Asking a single question but allowing students to choose from a variety of stems to construct responses. Students choose a stem based on their level of language knowledge and proficiency. (Seidlitz & Perryman, 2008)

Total Physical Response (TPR): A way of teaching using gesture and movement so that content is comprehensible to ESL newcomers. (Asher, J., 1967)

Total Response Signals (Also called active response signals): Active responses by students such as thumbs up/down, white boards, and response cards. Response signals enable teachers to instantly check for understanding and allow students to self assess current levels of understanding.

Unit Study for ELLs: A modified approach to writers workshop advocated by Samway (2006). The steps involve:

- Gathering high quality samples of the genre
- Immersion in the books
- Sifting between books that students can model and those that they can't

- Students immerse themselves a second time in the books
- Students try out using the "writing moves" they find the accomplished writers using
- Writing and publishing
- Reflecting and assessing

Visual Literacy Frames: A framework for improving visual literacy focusing on affective, compositional, and critical dimensions of processing visual information (Callow, J., 2008).

Visuals: Illustrations, graphic organizers, manipulatives, models, and real world objects used to make content comprehensible for English learners.

Vocabulary Alive: Students memorize a lesson's key vocabulary by applying gestures to each term. The gestures can be assigned by the teacher or by students. Once the gestures are determined, each term and its gesture is introduced by saying, "The word is _____ and it looks like this _____." (Created by Cristina Ferrari, Brownsville ISD)

Vocabulary Game Shows: Using games like Jeopardy, Pictionary, and Who Wants to be a Millionaire etc., to have students practice academic vocabulary.

Vocabulary Self Collection: A research based method of vocabulary instruction involving student collection of words for the class to study. Students share where the word was found, the definition and why the class should study that particular word. (Ruddell, M., & Shearer, B., 2002)

W.I.T. Questioning: A questioning strategy involving training the students to use three stems to promote elaboration in discussion (Seiditz & Perryman, 2008):

- Why do you think...?
- Is there another...?
- Tell me more about...

Whip Around: A way of getting input from all students during a class discussion. The teacher asks students to write a bulleted list in response to an open ended question. Students write their responses to the question and then stand up. The teacher then calls on students one at a time to respond to the question. If students have the same answer they mark it off on their papers. The teacher continues to call on students and students continue to mark through their answers. When all their answers have been marked through the students sit down. The activity continues until all students are seated. (Fisher, D. & Frey, N., 2007)

Word Analysis: Studying the parts, origins, and structures of words to improve spelling (Harrington, 1996).

Word Generation: Students brainstorm words having particular roots. Teachers then have students predict the meaning of the word based on the roots. (Echevarria, Vogt & Short, 2008)

Word MES Questioning: A method of differentiating instruction for ELLs developed by Hill & Flynn (2006). The mnemonic device stands for "Word, Model, Expand, and Sound." Teachers work on *word* selection with pre-production students. "*Model* for early production. *Expand* what speech emergence students have said or written and help intermediate and advanced fluency students *sound* like a book" by working on fluency.

Word Play: Manipulating words through various word games to increase understandings. Johnson, von Hoff Johnson, & Shlicting (2004) divide word games into eight categories: onomastics (name games), expressions, figures of speech, word associations, word formations, word manipulations, word games, and ambiguities.

Word Sorts: Sorting words based on structure and spelling to improve orthography (Bear, D. & Invernizzi, M., 2004).

Word Study Books: A way of organizing words into a notebook based on spelling and structures such as affixes and roots. (Bear, D., & Invernizzi, M., 2004).

Word Walls: A collection of words posted in a classroom organized by topic, sound, or spelling to improve literacy. (Eyraud et al., 2000)

This page is intentionally left blank.

Bibliography

Asher, J. and Price, B. (1967). The learning strategy of total physical response: Some age differences. *Child Development*, 38, 1219-1227.

Asher, J. (1969). The total physical response approach to second language learning. *The Modern Language Journal* (53) 1.

August, D. and Shanahan, T. (2006). *Developing literacy in second-language learners: report of the national literacy panel on language-minority children and youth.* Center for Applied Linguistics, Lawrence Erlbaum Associates: Mahwah, NJ.

Ausubel, D. P. (1960). *The use of advance organizers in the learning and retention of meaningful verbal material.* Journal of Educational Psychology, 51, 267-272.

Baker, L. (2004). Reading comprehension and science inquiry: Metacognitive connections. In E.W.Saul (Ed.), *Crossing borders in literacy and science instruction: Perspectives on theory and practice.* Newark, DE: International Reading Association; Arlington, VA: National Science Teachers Association (NSTA) Press.

Bauman, J. F., Russell, N. S., and Jones, L. A. (1992). Effects of think-aloud instruction on elementary students' comprehension abilities. *Journal of Reading Behavior*, 24 (2), 143-172.

Bear, D.R., Invernizzi, M., Templeton, S., & Johnson, F., (2004). *Words their way: Word study for phonics, vocabulary, and spelling instruction (2nd Ed.).* Upper Saddle River, NJ: Merrill Prentice Hall.

Beck, I.L., McKeown, M.G., Hamilton, R.L., & Kugan, L. (1997). *Questioning the author: An approach for enhancing student engagement with text.* Newark, DE: International Reading Association.

Berg, C. (1999). The effects of trained peer response on ESL students' revision types and writing quality. *Journal of Second Language Writing, Volume 8, Issue 3, September 1999, Pages 215-241.*

Boulware, B.J., & Crow, M. (2008, March). Using the concept attainment strategy to enhance reading comprehension. *The Reading Teacher, 61*(6), 491–495.

Brown, A., Campoine, J., and Day, J. (1981). Learning to learn: On training students to learn from texts. *Educational Researcher, 10,* 14-24.

Bruner, J., Goodnow, J. & Austin, G. A. (1967). *A study of thinking.* New York: Science Editions.

Chamot, A.U. & O'Malley, J.M. (1994) The calla handbook: implementing the cognitive academic language learning approach. White Plains, NY: Addison Wesley Longman.

Callow, J. (2008, May). Show me: principles for assessing students' visual literacy. *The Reading Teacher, 61*(8), 616–626.

Celce-Murcia, M., Brinton, D. & Goodwin, J. (1996). *Teaching pronunciation: A reference for teachers of English to speakers of other languages.* Cambridge: Cambridge University Press.

Cunningham-Flores, M. (1998) *Improving adult esl learners' pronunciation skills.* National Center for ESL Literacy Education.

Dole, J., Duffy, G., Roehler, L., & Pearson, P. (1991). Moving from the old to the new: Research in reading comprehension instruction. *Review of Educational Research, 61,* 239-264.

Echevarria, J., Short, D & Vogt, M. (2008). *Making content comprehensible. The sheltered instruction observation protocol.* Boston, MA: Pearson

Elbow, P. (1998) *Writing with power.* Oxford: Oxford University Press.

Eyraud, K., Giles, G., Koenig, S., & Stoller, F. (2000). The word wall approach: Promoting L2 vocabulary learning. *English teaching Forum*, 38, pp. 2-11.

Fisher, D., & Frey, N. (2007). *Checking for understanding: Formative assessment techniques for your classroom.* Alexandria, VA: Association for Supervision and Curriculum Development.

Francis, D., Lesaux, N., & August, D. (2006). Language of instruction for language minority learners. In D. L. August & T. Shanahan (Eds.) *Developing Literacy in a second language: Report of the National Literacy Panel*. (pp.365-414). Mahwah, NJ: Lawrence Erlbaum Associates. (2006)

Gibbons, P. (2002) *Scaffolding language, scaffolding learning.* Portsmouth, NH: Heinemann.

Goldenberg, C., (1992-1993) Instructional conversations: promoting comprehension through discussion, *The Reading Teacher, 46 (4),* 316-326.

Harrington, M. J. (1996). Basic instruction in word analysis skills to improve spelling

competence. *Education*, 117, 22. Available: http://www.questia.com/

Head, M., & Readence, J. (1986). *Anticipation guides: Meaning through prediction.* In E. Dishner, T. Bean, J. Readence, & D. Moore (Eds.), (1986) *Reading in the Content Areas,* Dubuque, IA: Kendall/Hunt.

High, Julie. (1993). *Second language learning through cooperative learning.* San Clemente, CA: Kagan Publishing.

Hill, J., & Flynn, K. (2006). *Classroom instruction that works with English language learners.* Alexandria, VA: Association for Supervision and Curriculum Development.

Johnson, D., & Johnson, R. (1995). *Creative controversy: Intellectual challenge in the classroom* (3rd ed.). Edina, MN: Interaction Book Company.

Kagan, S. (1990). *Cooperative learning for students limited in language proficiency.* in M. Brubacher, R. Payne & K. Rickett (Eds.), (1990) *Perspectives on small group learning.* Oakville, Ontario, Canada.

Kagan, S., (1992). *Cooperative learning.* San Juan Capistrano, CA: Kagan Cooperative Learning.

Kirschner, P., Sweller, J., & Clark, R. (2006). "Why minimal guidance during instruction does not work: an analysis of the failure of constructivist, discovery, problem-based, experiential, and inquiry-based teaching". *Educational Psychologist* 41 (2): 75–86.

Krashen, S. (1982). *Principles and practices in second language acquisition.* Oxford: Pergamon.

Leeman, J. (2003). Recasts and second language development: Beyond negative evidence. *Studies in Second Language Acquisition, 25,* 37-63.

Lyman, F. T. (1981). The responsive classroom discussion: The inclusion of all students. In A. Anderson (Ed.), Mainstreaming Digest (pp. 109-113). College Park: University of Maryland Press.

Marzano, R. (2004) *Building academic background.* Alexandria, VA: MCREL, ASCD.

Marzano, R., Pickering, D. J., & Pollock, J. E. (2001) *Classroom instruction that works.* Alexandria, VA: MCREL, ASCD

Maurice, K., (1983) The fluency workshop. *TESOL Newsletter,* 17, 4.

Mayer, R. (2003) *Learning and instruction.* New Jersey: Pearson Education, Inc.

National Council of Teachers of Mathematics (NCTM). (2000). Principles and standards for school mathematics. Reston, VA: NCTM.

Novak, J.D. (1995), *Concept mapping: a strategy for organizing knowledge.* in Glynn, S.M. & Duit, R. (eds.), *Learning Science in the Schools: Research Reforming Practice*, Lawrence Erlbaum Associates, (Mahwah), 1995.

Ogle, D. S. (1986). K-W-L group instructional strategy. In A. S. Palincsar, D. S. Ogle, B. F. Jones, & E. G. Carr (Eds.), *Teaching reading as thinking* (Teleconference Resource Guide, pp. 11-17). Alexandria, VA: Association for Supervision and Curriculum Development.

Palincsar, A.S., & Brown, A.L. (1985). Reciprocal teaching: Activities to promote reading with your mind. In T.L. Harris & E.J. Cooper (Eds.), *Reading, thinking and concept development: Strategies for the classroom*. New York: The College Board.

Paulk, W. *How to Study in College*. Boston: Houghton Mifflin, 2000.

Pilgreen, J. 2000. The SSR Handbook: How to Organize and Maintain a Sustained Silent Reading Program. Portsmouth, NH: Heinemann.

Pilgreen, J. and Krashen, S. 1993. Sustained silent reading with English as a second language with high school students: Impact on reading comprehension, reading frequency, and reading enjoyment. School Library Media Quarterly 22: 21-23.

Powell, M. 1996. *Presenting in English*. Hove: Language Teaching Publications.

Chamot, A., & O'Malley, J. (1994). *The calla handbook: Implementing the cognitive academic language learning approach*. Reading, MA: Addison-Wesley

Ruddell, M.R., & Shearer, B.A. (2002). "Extraordinary," "tremendous," exhilarating," "magnificent": Middle school at-risk students become avid word learners with the vocabulary-self collection strategy (VSS). *Journal of Adolescent and Adult Literacy, 45*(4), 352-363.

Samway, K. (2006). *When English language learners write: connecting research to practice.* Portsmouth: Heineman.

Schlick Noe, K. & Johnson, N., 1999). *Getting started with literature circles*. Norwood, MA: Christopher-Gordon Publishers, Inc.

Seidlitz, J. & Perryman, B., (2008) *Seven steps to building an interactive classroom: Engaging all students in academic conversation.* San Antonio TX: Canter Press.

Taba, H. (1962). *Curriculum development: Theory and practice.* New York: Harcourt Brace & World

Taba, Hilda. (1967) *Teachers' handbook for elementary social studies*. Reading, MA: Addison-Wesley.

Taylor, W. (1953). Close procedure a new tool for measuring readability. *Journalism Quarterly.* 30, 415-433.

Thornburry, S. (2005). *How to teach speaking.* Essex, England: Pearson.

Vail, Neil J. and Papenfuss, J. (1993). *Daily oral language plus.* Evanston, IL: McDougal, Littell.

Weaver, C. (1996). *Teaching grammar in context.* Portsmouth, NH: Boynton, Cook Publishers.

Wennerstrom, A. (1993). Content-based pronunciation. *TESOL Journal*, 1(3), 15-18.

White, T., Sowell, J., & Yanagihara, A. (1989). Teaching elementary students to use word-part clues. *The Reading Teacher, 42*, 302-308.

Willhelm., J (2002). *Action strategies for deepening comprehension*. New York: Scholastic.

Zwiers, J. (2008). *Building Academic Language*. Newark, DE: Jossey-Bass/International Reading Association.

Please visit **www.seidlitzeducation.com** for upcoming events, district trainings, and products.

Order Form

NEW!
RTI for ELLs: *Considerations for Success with Diverse Learners*
Suggested Guidelines for making RTI successful for ELLs.

NEW!
***'Instead of I Don't Know…'* Poster**
This classroom poster is an excellent resource for both teacher and student offering some helpful phrases to replace the all too common response, "I don't know."

Secondary>

< Elementary

ORDERING METHODS

- **ONLINE** at **seidlitzeducation.com** or
- **FAX** completed order form with payment information to **(210) 587-2495**

Product	*Price*	*Quantity*	*Total*
NEW! **ELPS Flip Book**	**$19.95**		
Navigating the ELPS: Using the New Standards to Improve Instruction for English Learners John Seidlitz	**$24.95**		
Navigating the ELPS: Math John Seidlitz and C. Araceli Avila	**$29.95**		
Navigating the ELPS: Science John Seidlitz and Jennifer Jordan-Kaszuba	**$29.95**		
Navigating the ELPS: Social Studies John Seidlitz and Bill Perryman	**$29.95**		
Navigating the ELPS: Language Arts and Reading John Seidlitz	**$34.95**		
Navigating the ELPS: Sentence Cubes: Math, Science, History, & ELA	**$20.00**		
NEW! 'INSTEAD OF I DON'T KNOW' Poster ☐ Elementary ☐ Secondary	**$9.95**		
NEW! RTI for ELLs Fold-Out	**$16.95**		

SHIPPING 1-5 books $10.95 | 6-10 books $13.95 | 10 or more books – call for rates.
5-7 business days to ship. If needed sooner please call for rates.

TAX EXEMPT? please fax a copy of your certificate along with order.

DISCOUNT	
SHIPPING	
TAX	
TOTAL	

NAME

SHIPPING ADDRESS   CITY   STATE, ZIP

PHONE NUMBER   EMAIL ADDRESS

***ORDERING By FAX to* (210) 587-2495**
please complete credit card info *or* attach purchase order

☐ **Visa** ☐ **MasterCard** ☐ **Discover** ☐ **AMEX**

CC#   Expiration Date:

Signature

☐ **Purchase Order attached**
please make P.O. out to **Encompass Event Planners***

To schedule training or for more information, please contact us at (210) 591-8650

10864 Gulfdale St.
San Antonio, TX 78216
www.seidlitzeducation.com

Pricing, specifications, and availability subject to change without notice.

*Encompass Event Planners is the sole source distributor for Seidlitz Education